←

This pocket and the one inside the back cover are intended to hold the two series of guidelines—one series of booklets for special language groups and the other for the different disciplines of science—that are being prepared and issued separately from this volume. (Please note that these booklets will *not* be published by Associated Scientific Publishers; further information about the two series can be obtained from the Secretary of ELSE: Dr. J. R. Metcalfe, Commonwealth Institute of Entomology, 56, Queen's Gate, London, SW7 5JR, UK.)

Writing Scientific Papers in English

ELSE (European Life Science Editors) is the acronym adopted by the European Association of Editors of Biological Periodicals.

The Ciba Foundation, London, is an educational and scientific charity for the promotion of international cooperation in medical and chemical research.

EDITERRA is the European Association of Earth-Science Editors.

Writing Scientific Papers in English

An ELSE - Ciba Foundation Guide for Authors

Maeve O'Connor
Senior Editor, Ciba Foundation, London

and

F. Peter Woodford
Executive Director, Institute for Research into Mental & Multiple Handicap, London

Sponsored by the International Union of Biological Sciences, the International Union of Geological Sciences, and EDITERRA

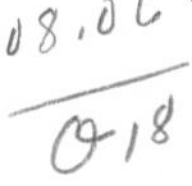

1976

Elsevier · Excerpta Medica · North-Holland

Amsterdam · Oxford · New York

First printing January 1975
Second printing August 1975
Third printing January 1976
Fourth printing March 1976

ISBN Excerpta Medica 90 219 4035 3
ISBN American Elsevier 0444-15165-6

Published in 1976 by Elsevier/Excerpta Medica/North-Holland, P.O. Box 211, Amsterdam, and American Elsevier, 52 Vanderbilt Avenue, New York, N.Y. 10017.

Suggested publisher's entry for library catalogues: Elsevier/Excerpta Medica/North-Holland.

Printed in The Netherlands by Trio, The Hague

Foreword and acknowledgements

The inception of this book lies some way back in the history of ELSE (the European Association of Editors of Biological Periodicals). English has taken the place once held by Latin in the world of learning, and it was natural that the problems of how to formulate the results of a scientific enquiry in English should become a target for this multilingual group of editors; at a meeting of the association in Amsterdam in 1967 a Working Group was established, and preliminary studies were initiated. In 1973, at the 2nd General Assembly of ELSE held in Ustaoset, Norway, detailed suggestions for a European Style Manual were presented and discussed, and a Style Manual Committee was appointed by Ernest Mann, President of ELSE.

The Style Manual now published as a 'Guide for Authors' was envisaged as forming the nucleus or core of a system of specific guidelines for special language groups and for the different disciplines of science. The difficulties encountered by Scandinavians trying to write English are different from those encountered by Russians, and physiologists face problems different from those that concern taxonomists. Those problems will be dealt with in a series of small companion booklets which are being prepared and distributed separately from this 'core manual'. (The pockets inside the cover of the volume have been designed to hold those booklets as you obtain them.)

The ELSE Style Manual Committee was fortunate in obtaining the active support of the Ciba Foundation and of its Director, Gordon Wolstenholme, and to find in Maeve O'Connor and Peter Woodford two authors who brought a rare combination of enthusiasm, knowledge and drive to their work. John Metcalfe, the Secretary of ELSE, was another valuable member of the team that produced this book. Arthur Simpson, Chairman of the Handbook Committee of EDITERRA (the European Association of Earth-Science Editors) also lent the support of his Committee. I therefore want to thank everyone who contributed actively, those who presented their views before and after the Ustaoset meeting as well as those who later commented on the various drafts of this book, among whom are M. Balaban, D. N. Baron, R. Borth, J. Bureš, C. Findell, D. FitzSimons, J. W. Glen, J. Gravesteijn, M. H. Hey, E. J. Huth, J. King, J. Knight, G. Lea, G. Mathé, F. B. O'Connor, J. C. Rigg, H. Samulowitz, L. Walschot, C. F. Winkler Prins, S. Woodford and E. Zaimis.

KNUT FÆGRI

Chairman, ELSE Style Manual Committee

Contents

Introduction

Getting down to writing a paper for a scientific journal is like trying to start an old car on a frosty morning: the would-be driver is anxious, the car is cold and reluctant, and both man and machine suffer for a while. In this book we try to help authors through this painful period by suggesting how they can get the machinery running smoothly. Editors and referees will sometimes rescue an author whose prose has broken down—but they are busy people whose humanitarian instincts should not be abused, and it is better for all concerned if authors try to submit papers that are in good working order.

This guide is for scientists of any nationality who want to submit papers to journals published in English. The difficulties of preparing a paper are naturally greater for those who have to express themselves in a language not their own. But authors should neither worry too much about language difficulties nor use them as an excuse for failing to unravel tangled lines of thought. The most troublesome blocks to writing a good scientific paper are conceptual and procedural. When underlying failures in logic have been dealt with, many language problems either disappear or can easily be solved for the author by someone else with a better command of language. We have therefore *not* attempted to give a course in English grammar here, though chapter 4 deals briefly with some errors often made in scientific writing.

Although this book is mainly for inexperienced writers and those who write with difficulty, it should also help scientists of greater experience who want editors to accept their papers more readily and who want readers to understand, remember and even enjoy those papers. In addition, we have tried to help the typist (whether the typist is the author or someone else) who will be preparing the typescript for submission to a journal.

The advice given here is directed towards the writing of original scientific articles (excluding review articles) for journals. However, the basic principles are valid for any kind of scientific writing and can be applied, with suitable modifications, to theses or dissertations, technical reports, research project proposals and chapters for books.

Appendix 1 lists a sequence of steps in writing a paper. Readers who do not like the sequence we describe will be able to use Appendix 1 (which is in fact the outline of the book) to devise a sequence which suits them better.

Our guidelines are in general agreement with the recommendations to authors contained in the (American) *CBE Style Manual* (Council of Biology Editors 1972), but they take both British and American English usage into account as well as certain European printing practices that differ from those in the USA. Likewise, the guide is consistent with the CBE manual entitled *Scientific Writing for Graduate Students* (Woodford 1968), but differs from it in being addressed to authors rather than to teachers of authors. The guide is also compatible with the relevant British, American and international standards listed in the bibliography (pp. 99–101), with the *General Notes on the Preparation of Scientific Papers* produced by the Royal Society (1974), and with the draft *Guide to Authors* prepared for Australian journals published by CSIRO (1973).

The recommendations made here do not constitute a set of rules to be strictly obeyed in all circumstances. The most relevant guide for an author is the latest set of Instructions to Authors issued by the journal in which he or she hopes to publish. Those instructions must be given priority over any suggestions we put forward. We hope, however, that journals will increasingly adopt recommendations which reflect—as we believe ours do—the standards accepted by most of the scientific community.

Chapter 1

Planning

What should you submit for publication in a scientific journal, when should you submit it, and where? You must deal with these questions before you think about *how* to write a paper. Finding satisfactory answers before you begin to write will save your own time and the editor's—and the paper that you write will stand a better chance of appearing in a first-rate journal.

What to Write and When

Scientific papers should describe significant experimental, theoretical or observational extensions of knowledge, or advances in the practical application of known principles. What is the best way of deciding whether what you have to report adds significantly to present knowledge, understanding or practice? Our advice is: *begin at the end.* Write down your conclusions as clearly, precisely and economically as you can, and relate them to the question or hypothesis you have been examining. (Note that the question or hypothesis need not be the one with which you started the investigation. A scientific paper should not be the history of an enquiry but its outcome.) Ask yourself how tentative or firm the conclusions are, measure them against previous knowledge of the subject, and discuss them, if you can, with an experienced scientist from another field. Listen carefully to that scientist's general assessment of what you propose to report and to his or her questions about your assumptions, evidence and lines of reasoning. Then decide whether you are really ready to publish, and if so whether what you have to say will be more suitable for a long article or a short one, for a brief note or for a preliminary communication. Severe self-criticism at this stage will save you from more painful criticism by others later.

This process of self-criticism may seem unnecessary, especially if you are sure you have reached the end of an investigation, but it is in fact essential to success in writing a paper. In doing what we suggest you may find that you are indeed at the end of a road, but a road different from the one on which you started out a year or two ago. You must recognize this before you prepare to write the full paper, otherwise the thoughts underlying your writing will be muddled. (So useful is this technique of writing to clarify thinking that many people write a long essay to themselves halfway through an investigation to reassure themselves that they are on the right lines and to clarify what remains to be done.)

What not to write

Do not write an article simply because you think it is time your name appeared in print; and do not try to spin several short papers out of what is essentially one investigation. If you have developed a new method and then applied it to a problem, two papers (one on the method, another on the results) *may* be justified if the method is based on a new principle or has wide applicability. Similarly, when a large-scale investigation obviously has to continue for many years, or falls naturally into several sections complete in themselves, publication of several papers may be appropriate. But never publish for self-glorification—your motives will be only too obvious to the experienced reader.

Before you go further in planning a paper, consider whether official secrecy acts will form a barrier to publication. You may need to take legal advice on this. Similarly, publication of an article describing a new method or machine may invalidate an application for a patent.

Need for originality

A scientific paper worth submitting to a journal must describe previously unpublished work. (A *review* article will, of course, discuss previously published scientific work; its originality lies in the discriminating selection of material for comment and in the author's assessment of the current state of research on the topic under review.) Never, therefore, submit for publication a paper that has already been published or accepted for publication elsewhere, or one that is being considered for publication. Further, do not write up the same results in different forms in the hope of squeezing more than one publication out of them. Duplicate publication damages the writer's reputation; it is also unpopular both with journals and with documentation and abstracting services, chiefly because of the high costs of assessing, editing and printing the articles, of publishing and indexing the abstracts, and later of retrieving the information contained in the article.

The originality of a paper is not invalidated by previous publication of a *preliminary communication*, provided the latter is brief and lacks detail, and provided it is referred to as such in the subsequent paper. But remember that preliminary communications have the drawback that if they appear long before the detailed version they may produce misunderstandings or lead to unwarranted generalizations. The published abstract of a talk at a scientific meeting will not put an article describing the same results into the 'previously published' category. But if a thesis or dissertation has been widely distributed (for example, through microform sales), or if a detailed account of the work has appeared in a scientific or general newspaper, the editor of a journal may well reject an article based on the same work. Very rarely, the same article may be published in journals addressed to different audiences (for example, general practitioners and librarians), or in different languages, or for very different geographical distributions; but even fully justifiable duplicate or multiple publication is allowable only with the full knowledge and consent of both or all of the journal editors concerned.

Where to Submit a Paper

Now that you have an idea of what sort of paper you want to write, the next question is: which journal should you write it for? Do not postpone this decision to a later stage. All writers, including scientists, have to keep their readers in mind, and writing for a particular journal will enable you to imagine your future readers more clearly. Conversely, it is unwise to prepare an article for a journal whose editor is certain to reject it because the subject matter does not fall within the journal's scope.

Be careful about submitting papers to new journals. Some are started because the chief editor thinks it would add to his prestige or because the publisher wants to extend his list. Such journals often have a lower standard than established ones because they need to accept virtually everything submitted in order to fill their pages. If that state continues, the journal may not last more than a year or two and your work may easily be overlooked.

In selecting your 'target journal', ask yourself whether the work you want to publish is of wide general interest, of interest to most people in your discipline, or of specialized interest to a few. The answer will narrow the field of possible journals considerably, with any luck down to very few. Make your final choice among them by reading their own statements of purpose and scope. Such statements appear in most journals at the beginning of the notes (or instructions) to authors.

If several journals appear suitable, look at each of them critically from several points of view. Examine the scientific quality of the articles published.

See how well the illustrations are reproduced. Find out which of the journals have a good record for thorough and fair editorial review and helpful refereeing. Determine how quickly, on average, each of the journals under consideration publishes articles after receiving them; whether it circulates widely or sells only a few hundred copies; and whether full abstracts of the articles published appear in a leading abstract journal within a year or so. *Ulrich's International Periodicals Directory* will give you the last two pieces of information, but only experienced authors can assess the standard of refereeing for you. A list of distinguished scientists on a journal's editorial or advisory board is not—unfortunately—any guarantee of a good editorial system or careful refereeing. An eminent scientist named as editor-in-chief should, however, influence your choice, because he bears responsibility for the journal's contents, and his reputation will suffer if the standard is not high.

Obtaining the Instructions to Authors

Even if you decide to submit your paper to a journal you often read and with which you are completely familiar, so that you know what subjects it covers and the kind of articles it publishes, obtain and read the Instructions to Authors before you begin preparing your article. Some journals include these instructions in every issue, some publish them only in the first issue of each volume, and some supply them only if you write to the editorial office. Since editorial policy and practices evolve continuously, be sure to consult the most recent set of instructions. Reading them should enable you to decide, finally, whether the journal is the most suitable one for the paper you are going to write. Even though some of the instructions may seem too detailed to consider yet, read them through quickly to learn the journal's recommended length for articles, the preferred order of main sections, special language requirements, recommended dictionaries or style guides, and policy on abbreviations and nomenclature, all of which you need to know as you go on to prepare your article.

Chapter 2

Preparing

Deciding on Authorship

You have assessed your work, decided that it is worth submitting to an editor, and chosen the most suitable journal. Unless you worked alone, the next step is to decide whose names will appear as authors of the paper. Agree on this now with your colleagues and prevent later arguments. Include as authors all those, but only those, who contributed substantially to the theoretical or experimental work. Although it is better if just one person does the writing, all those named as authors must read and approve *at least* the final draft before the paper is submitted. Do not include as authors people who simply advised you or gave technical assistance in the normal course of their work. Do not automatically list the head of your department or other senior colleagues unless they contributed significantly to the work you are reporting. Never include anybody as an author without obtaining his or her approval.

Establish also the order in which the authors' names will be given. There are no binding, universally agreed rules about the order for names. If one of the authors unquestionably contributed most to the work being reported, put his or her name first, followed by the others in an agreed order reflecting how much each person did; if everyone contributed equally, arrange the names alphabetically or according to the conventions of your country and laboratory. Members of teams who write several papers together often take turns at being named as first author. If one person is clearly the leader of the research team, or if the head of the department or another senior colleague has contributed to the work, the local convention may be that that person's name always appears either first or (more often) last.

In what follows, we make the assumption that you are the author who is doing most of the work of preparing the paper, with assistance and criticism from your co-authors. If, as is often done, you plan to divide the writing between a more experienced and a less experienced writer, this chapter and chapters 3–5 will help you to share out the work appropriately.

Determining the Journal's Requirements

Begin by considering any special requirements of the journal to which you intend to submit your paper (referred to here as 'the target journal' or 'the journal'). If the Instructions to Authors specify certain style guides or rules of nomenclature, obtain and study these without delay.

If your work includes experiments on human beings or animals, find out what the editor wants you to report concerning your adherence to ethical principles and guidelines, especially if the journal is a foreign one. The basic ethical principles for research investigations in man and other animals are universal, but attitudes towards procedural details are changing in different countries at different rates. Although your experiments may not conflict with local practices, your article describing them may be unacceptable to a journal in a country with different regulations.

Plan to use the approved international system for units of measure (Système International d'Unités, SI) unless the journal expressly forbids it. This system was adopted by the Conférence Générale des Poids et Mesures as long ago as 1960 and it is a coherent, unambiguous and internationally accepted system applicable to all disciplines. It is described in Appendix 2, which includes conversion factors from non-standard units. If you have made measurements in other units, convert the values to SI units at this stage. You may occasionally add values in traditional units (in parentheses) if you think this will help some readers.

Composing the Working Title and Abstract

Next compose a working title and a draft abstract (that is, a short version of what you intend to say) for the paper you have not yet written. This suggestion is not as paradoxical as it may seem. You have already worked out what your conclusions are and how they relate to other work on the same subject. You have recorded your observations and made notes on what you have read. Drafting the title and abstract should therefore present no difficulties, if you know exactly what your article is to convey. If you are not yet sure of the scope of your article, composing a working title and abstract will help you to define it. Later, the working title and abstract will guide you in deciding what to include and what to leave out of the article. Refer back to them often while you are preparing and writing the paper.

The working title can be as long or as short as you like, but it must contain only one topic, which will be the main subject and point of your paper. In the working abstract, express concisely your hypothesis, the approach you adopted, your conclusions, and your assessment of their significance. Try to do this in four sentences: one for each item.

The *final* title and abstract (see pp. 46–49) will play an important part in guiding the reader and in information retrieval, but the *working* title and abstract have quite a different purpose, namely to help you, the author, to put your ideas together in an orderly form.

Organizing the Paper

Now decide the basic structure of the paper. Read the Instructions to Authors again, or examine the papers in a few issues of the journal to determine the structure most commonly used in that journal. In biomedical journals the conventional division into:

Introduction
Materials and Methods
Results
Discussion

has great value. Possible headings for a paper in a descriptive field science might be:

Introduction
Materials and Methods
Geographical Context
Analysis of Data
Results
Discussion
Conclusions

Headings for a theoretical paper might be:

Introduction
Theoretical Analysis
Applications
Conclusions

and in a paper describing a new method:

Introduction
Description of Procedure
Tests of New Method
Discussion

You need good reasons for modifying or abandoning one of the conventional patterns in your field, but you must certainly do so if your material demands it. For instance, results can sometimes be effectively combined with discussion; again, it may be a good thing to give some results first, in order to explain why you adopted the methods which you then go on to describe. Procedural details may sometimes be placed in an appendix.

If the journal requires or favours a section headed Conclusions, or Conclusions and Summary, take this into account in planning the Discussion section which precedes it.

If you want to include detailed proofs or bulky data not essential to the main theme, consider whether you will need an Appendix. An Appendix is not a note added in proof; it should form part of your original plan for the paper, and it will undergo editorial review with the rest of the paper.

Collecting the Material

When you have selected the main (i.e. *first-order*) headings as just discussed, write each of them at the top of a sheet of paper. On each sheet put down in random order, just as they occur to you, all the findings or thoughts that seem to belong under that heading. You may use a personal shorthand instead of sentences. Look through your laboratory notes and records, and list everything that is relevant to the topic defined by the working title—but nothing that is not. Add to the relevant section any descriptions of materials, methods or results that you have already written. Take care to assign each item to the correct section as defined by the heading. If an item could be assigned to more than one section, decide which is the more appropriate or make a note that this decision will have to be made later. Do *not* try to assemble the items in logical order yet.

As you go along, add any relevant tables, graphs or other illustrations that you have already made in rough form. Collect also all the reprints or cards giving bibliographical details of references that you are likely to need.

Designing Tables and Illustrations

Tables and illustrations usually contain all the evidence on which a paper is based, and readers often examine them before deciding whether the text is worth studying. Choosing and designing them carefully now, before you write the text, will show up any gaps in your evidence and tell you whether more experiments or observations are needed or whether you should modify (or even abandon!) your conclusions. It will also ensure that a reader will be able to grasp the significance of tables and illustrations without referring to the text. Finally, it may save you writing many extra words, because information presented in tables or illustrations need not and should not be repeated in detail in the text.

Every illustration and table must have a purpose and form an essential part of your argument. Aim to make each of them convey a clear message, whether it be a definitive finding or a trend in values. This means that you must pay close attention to the title and legend as well as to the design of the non-verbal part of the illustration. Try to provide maximum information in minimum space, while keeping the whole as simple and clear as possible.

When you can choose between presenting results in a table or a graph, base your choice on the fact that tables can more accurately present numerical results for comparison with work reported elsewhere, while graphs reveal trends and relationships more vividly.

Keep a copy of the journal and its Instructions to Authors in front of you when you are designing and constructing the tables and illustrations, and check whether there is any limit to their number or size.

Tables

Model the shape, size and framework of the tables on the style used in your target journal. The titles, column headings, entries in the first column and explanatory notes form the framework of the tables. Together they must make the tables comprehensible and independent of the text. Make sure that the proposed tables are all worth setting as tables: very small tables can often be converted into one or two lines of text.

Number the proposed tables in the order in which you expect them to appear in the paper; use arabic numerals (1, 2, 3 . . .) unless the journal insists on roman numerals (I, II, III . . .). Then write a list of tentative titles, stating in each title what the table will show (conclusions as well as numbers) and perhaps indicating the experimental design, but keeping the titles short (for example, 'Table 1. Increased concentrations of aromatic amino acids in insects under zero-gravity conditions. Table 2. Decreased total amounts of aromatic amino acids in rat brains under zero-gravity conditions.'). Listing the titles together in this way will help you to decide which tables are essential and whether their purpose will be clear to the reader.

Compare the titles with one another and remove redundant words and inconsistent statements.

Within each table, arrange the columns in the order that will allow readers to understand most easily what the table is about and the conclusion(s) you want them to draw. For example, put columns that have to be compared beside each other. Put control or normal values near the beginning, perhaps in the second column or as the top row of figures. Arrange the numbers from small to large if it is logical to do this, or group the findings in some other logical way. Do *not* blindly list items in chronological order of experiment, or according to the page number in your notebook or alphabetically according to patients' initials.

In the body of the table, do not give numbers that imply greater accuracy than was attained. In biological work, four or more digits in a measured value are hardly ever justified. Give numbers to the nearest significant figure, rounding up or down as necessary; when the last number is 5, round off to the nearest even number. Eliminate surplus zeros from very large or small numbers by choosing the factors or units in the column headings carefully, to keep numbers between 0 and 1000 (e.g. convert 509 000 micromoles into 509 millimoles and write '509' in a column headed 'mmol'). Avoid numbers with exponents (10^6, 10^{-3}) as factors in headings; instead, convert the data into the appropriate units (milli, micro, nano and so on, as listed in Appendix 2) and write mg, μm, nl and so on at the top of the column. For dimensionless values, put exponents in the body of the table to avoid all ambiguity.

Avoid putting dashes in a column: they are ambiguous. Enter a zero reading as 0; insert an asterisk (*) or other symbol if an item was not measured, or if you tried but failed to obtain a value, and explain the sign in a footnote.

Where appropriate, state which test of significance you used, give the probability (P) values and the standard deviations or standard errors of the means (making it clear which), and *always* state on how many observations the values are based. An entry such as 6.0 ± 3.7, without explanation, does not convey accurate information.

For referring to explanatory notes to the table, use small superscript letters or numbers (^{a}, ^{b}; ^{1}, ^{2}), or symbols approved by the journal, in the body of the table. Use explanatory notes to tables to explain all uncommon abbreviations and for giving brief details of methods. If the same methods were used for work represented in several tables, put the necessary information in the explanatory notes to the first such table, and in the others refer the reader back to the first table for the details.

Make sure that the size and shape of the tables are suitable for your target journal before you decide on the final versions. Design tables narrow enough to fit either a column or a page of the journal. Editors dislike—but may accept—tables that have to be printed sideways on the page; they hate—and may ask you to shorten—tables stretching over two or more journal pages; and they or their printers will nearly always reject any tables necessitating a fold-out page. Many editors also dislike very narrow tables, unless two can be printed side by side. Sometimes, editors may ask for complicated tables to be prepared in such a way that they can be photographed and printed as illustrations.

Estimate the width of your draft tables by comparing the number of characters and spaces in the longest line of each table with the number in a printed line of a table in the journal. If a draft table seems too wide, decide whether all the columns are really necessary. Omit any column that can easily be calculated or deduced from the data in another column, unless you intend to discuss those particular values in detail. Can you leave out any words that are repeated in the left-hand column or the column headings? Can you transfer a column of reference values or of 'Notes' to the heading or the footnotes below the table?

If these tactics fail, try splitting an over-large table into two smaller ones, first asking yourself whether you really need to include every single value you obtained. Large collections of results or peripheral observations of interest to only a few readers can, on the editor's recommendation, be stored with a depository or retrieval centre (sometimes named in a journal's Instructions to Authors). However, you should always include enough values in tables to allow anyone repeating your experiments or basing further work on them to check their results against yours satisfactorily.

When each table matches the journal's requirements as closely as possible, look at all of them together: they should add up to a coherent story. If they do, tidy up the draft versions and keep them by you while you write the paper. Inconsistency between tables and text is perhaps the most frequent fault of scientific papers, and surely an easily avoided one.

Illustrations

Like the tables, the illustrations and their captions or legends must form a separate unit, independent of the text and fully self-explanatory. Your task at this stage is to select and design the illustrations and their legends.

Design the illustrations in consultation with a technical artist or photographer if possible, so that what is intellectually sound will later prove technically feasible. But do not have the illustrations made in final form yet, because this is expensive and time-consuming, and you may later decide to change or omit some of them.

There are two main kinds of illustration: *line drawings* (diagrams, histograms, graphs) and *photographs* from which half-tone illustrations can be made (for example, pictures of organisms, light or electron micrographs, photographs of instrumental tracings or of equipment). Good photographs of line drawings will sometimes be accepted, but they are difficult to correct; if you submit any, you will need to make quite sure that every detail on the original drawing is prepared exactly in accordance with the journal's requirements and that the photograph is in focus. Reprographed versions of illustrations are rarely acceptable. Do not plan to include line drawings and half-tone photographs in the same illustration unless you are sure that the journal is printed by the offset method (ask the editor if necessary).

In designing your line drawings do not fall into the habit of automatically showing numerical results in the form of a plotted curve; a histogram is often more appropriate, especially if there are no grounds for assuming the existence of a continuum between the experimental points.

You can use well-designed graphs (curves or histograms) to show several trends and relationships at once, but do not try to crowd too much into one figure. Show the standard deviation or the standard error of the mean for each point on a graph, where appropriate, by drawing a vertical bar; in the legend tell readers what the bar represents, and on how many observations each mean is based. Never extrapolate a line or curve beyond the observed points without reminding the reader of the hazards of drawing conclusions from the extrapolation.

Design illustrations so that as little lettering as possible will be needed. Draft short but informative descriptions for *x* and *y* axes (see Figs. 5 and 6, p. 35) and always give the units of measurement (using the Système International). Plan to use the same symbols for the same entities if they occur in several illustrations. Use the same coordinate system if values in different illustrations have to be compared.

When your impulse is to include photographs, consider first whether this is is in fact the best way of presenting your observations. Electrophoretograms, autoradiograms, schlieren records, paper chromatograms and similar illustrations usually reproduce poorly, and a short description often gives all the necessary information just as effectively and more economically. Alternatively, it may be more satisfactory to trace or copy line drawings from pen recordings and from chromatograms, etc. than to present photographs of the records themselves. A diagram or description of a new piece of equipment is often better than a photograph.

When you consider it essential to show the actual appearance of a record or object, choose or make photographs of the highest possible quality. They must be sharply focused, with good contrast between light and dark tones. Do not expect a printed illustration to turn out better than its original: in printing, the image is transferred from one surface to another two or three times, losing a little in clarity each time.

Consult the editor before you plan to include any colour illustrations. Many journals do not accept them; others may print them if you pay some or all of the very high cost. However, black and white prints can be made from colour slides, although there is some loss of definition.

Editors prefer illustrations that when reduced will fit a column or page of the journal and that use the available space economically: it is cheaper to make one large block for printing several illustrations together than to make several small blocks; it is also cheaper for the journal if all the photographs are rectangles. You will have to design, select, cut or mask illustrations so that their shape is suitable for your particular journal.

When you have designed the line drawings and selected the photographs, number the illustrations (or rough versions of them) in the order in which you expect to refer to them in the text. Do *not* number the photographs as a separate series from the drawings unless this is the journal's standard practice.

Now draft a set of legends for your illustrations. A legend consists of an informative title together with any necessary explanation of the illustration or the lettering or symbols included in it. It should not be placed on the illustration itself. Write all the legends in the style favoured by the journal, then type them as a consecutive series and keep them before you—together with the illustrations in their preliminary form—while you write the text of your paper. Do not make the final drawings or prints until you start revising the text (chapter 4).

Dealing with Copyright

So far, we have assumed that the illustrations and tables are all based on your own work. But occasionally you may want to borrow such items, as well as portions of text, from other people's published or unpublished work. Common politeness requires that you acknowledge the source of this material, including paraphrased passages. In many instances copyright law further requires you to obtain written permission to reproduce the material. It is your responsibility, not the editor's, to obtain that permission.

Copyright law is not always clear on how much you may quote from someone else's text without permission. It is safest to ask permission to quote anything longer than two or three lines, or about 50 words. If you want to reproduce a complete table or figure from published work, you must normally obtain written permission to do so from the copyright-holder—who may be the publisher, the author, or a third party named in one of the early pages of the book or journal in which the item first appeared. If the author of the item is not the copyright-holder, it is common courtesy, though not required by law, to obtain his approval as well as that of the copyright-holder. Asking for such approval may also be scientifically valuable, as the author may provide new information necessary to bring the quotation up to date. This is why we are suggesting that you write for permission at this early stage of preparing the paper.

Even when you wish to reproduce something from an earlier publication of your own, if you have not retained the copyright you must ask permission from the copyright-holder.

You can help to ensure a prompt reply to your request to reproduce material if you send both the copyright-holder and the author whose material you want to borrow *two* copies of a letter or release form such as the one shown opposite. If the form of acknowledgement is not specified in the reply, write 'Reproduced, with permission, from Smith 1973' and include the full reference in the list of references cited. Reproduction must be completely faithful to the original; any corrections or interpolations of your own must be identified as such (see also p. 36).

Where you have redrawn or rearranged a figure or table, write 'Based on Smith 1973' or 'Redrawn from Smith 1973', and include the full reference in the list of references cited.

If you quote from or refer to *unpublished* work (for example, work described in a letter or other informal document or in a conversation; or presented during a lecture or meeting that is not being published; or contained in an agency report of limited circulation), it is still more important to ask permission to reproduce or even to refer to it, although there may be no legal requirement to do so. The author or corporate author (an organization) may have strong objections to the work being released in this way or at the time you expect to publish; you may have misunderstood the author or he may have made a mistake which he can now correct; or subsequent work may have invalidated the original results provided. References to unpublished work are often textual rather than tabular, and we shall return to this point later (p. 26).

Outlining the Paper

Now, before you begin to write the text of the paper, make an outline. Two kinds of outline are commonly used: a topic outline and a sentence outline. A *topic outline* defines what subject will be discussed in each section or paragraph. A *sentence outline* expands the topic outline and puts into sentences the main point you want to make about each subject in the topic outline. Thus, the topic outline is a series of nouns or phrases arranged in a hierarchy, whereas the sentence outline is a series of sentences. Only the latter would make sense to another person. Ideally, every sentence in the sentence outline will form the basis of a paragraph in the final text.

Dear......

I am preparing an article entitled 'Going to the moon' for submission to Scientific Stargazer. I should be grateful if you would grant permission for the following to be reproduced:

Figs. 1 and 2 from your paper/the paper by I.M. Smith entitled 'Holes in space', published in Extraterrestrial Travel, 2, 99-101, 1973.

I am also writing separately to the copyright-holder/author requesting permission to reproduce this material. The usual acknowledgements and a full reference to the paper will of course be included. If you would like the credit line to take any special form, please let me know what this should be.

Would you please indicate your agreement by signing and returning one copy of this letter?
Thank you for your cooperation.

Yours sincerely,

O.H. Evans

I/we grant permission to reproduce the material specified above.

Signed:......................
(copyright-holder/author)

Date:........................

Credit line to be used:

....

(This sample letter is not copyright.)

Construct the topic outline by arranging the notes collected under your section headings in the order that seems most logical. Distinguish what is important or essential from what is subsidiary, supplementary or dispensable, and plan the relative amounts of space to be devoted to each topic accordingly. Arrange the topics in a hierarchy, on preferably not more than three or four levels. If necessary, assign symbols to these levels: I, II, III, etc. for the first level; A, B, C, etc. for the second; 1, 2, 3, etc. for the third; and *a*, *b*, *c*, etc. for the fourth. Then compose a statement about each topic and thereby form your sentence outline. Topics in a topic outline may be used later as the headings and subheadings of your first draft.

We give below an example of this method of working.

Example of the Method of Working Suggested Here

The unarranged notes we collected under the heading 'Chapter 2: Preparing' in preparation for writing this chapter looked like this:

Write draft title and abstract
SI
Co-authors
Read Instructions to Authors
Decide basic form
Copyright
Collect material without sequencing
Design tables and figures (colour)
Make outline
Ethics
Index terms
Permission to reproduce

After these had been rearranged in what seemed a logical order, the topic outline looked like this:

- Authorship
  - Inclusion
  - Order of names
- Journal's requirements
  - Instructions to Authors
    - Ethics
  - SI
- The draft title and abstract
- Basic form
- Gross organization of material
- Illustrations
  - Tables
  - Figures

Copyright
  Published
  Unpublished
Outlines
  Topic
  Sentence

From this we derived the following sentence outline: 'Decide on number and order of authors. Re-read the Instructions to Authors with principles of ethical experimentation and SI units in mind. Draft a tentative (working) title and abstract. Decide whether you will follow the conventional order of introduction, materials and methods, results, discussion. Jot down, in any order, all the items you will include under your chosen headings. Design the tables and figures, and write their titles and legends. Obtain permission to reproduce them when necessary, and consider the general question of copyright and the proper use of other people's work. Write a topic and a sentence outline of the paper.'

This outline includes nearly all the items listed for inclusion, rearranged in a logical order, but it does not correspond to an 'ideal' sentence outline, with every sentence eventually turning into a paragraph. The finished chapter, for example, has many paragraphs about designing tables and figures that received no specific mention in the outline; conversely, the item 'index terms' does not appear in this chapter (we decided to defer it until later). This illustrates the principle that you should write in accordance with the demands of the subject matter, and not follow the outline slavishly.

As you write more articles, you will discover (if you do not already know) whether a topic or a sentence outline best suits your needs. Some people prefer a *mixed* topic and sentence outline, in which each sentence is followed by a number of topics (not elaborated into sentences) that will be included in the paragraph eventually developed from the sentence. Whatever kind of outline(s) you prefer, it (or they) will show you both the form your paper is taking and whether there are any gaps in the logic.

When your outline is complete, with no gaps or repetitions, you are ready for the next stage: writing the first draft of the text.

Chapter 3

Writing the first draft

Once you have drafted a working title and abstract, drawn up the tables and illustrations, and written the outline(s), your article is well along the road to publication. Writing the text should present few difficulties if your preparations have been thorough.

Write the First Draft Quickly

Settle down at a time and place which will allow you to remain undisturbed for several hours. Collect all the material you have prepared and begin to write or type the first draft, or dictate it if that is easier for you. Follow the outlines closely at this stage, and write as quickly as you can. If the paper is short, try to finish the draft at one sitting so that it reads like a single unit rather than a series of unconnected passages. Write simply, without worrying about style or grammar. Do, however, think of the reader, and direct your thoughts and words not exclusively to fellow specialists, but to a wider audience of 'moderate specialists' (Royal Society 1974), particularly in the introductory and concluding sections. Further notes on these and the intervening sections are given below (pp. 21–24).

No matter what style the journal uses for citing references in the text (see pp. 52–53), use the name-and-date system ('Smith 1971; Braun *et al.* 1972') in writing the draft. Your final reference list will be more accurately compiled from names than from numbers that have to be changed as references are added or removed. If the target journal uses a numbering system, change the names to numbers only after you have prepared the *final* draft in the last stage of polishing the text (see pp. 54–57).

Although details of grammar can wait till later, try to use the appropriate person, tense, and voice of verbs in writing the first draft of the different sections.

Use 'I' or 'we' for describing what you did, 'you' or the imperative ('Put the coverslip on the slide . . .') for instructions, and the third person for describing what happened. The past tense is best for observations, completed actions and specific conclusions; the present is correct for generalizations and statements of general validity. Prefer the active voice ('I removed the needle') to the passive ('The needle was removed').

You have already selected your main (first-order) headings (p. 10). Insert subheadings and sub-subheadings freely as signposts for the reader and for yourself, especially if the paper is long. You can use the topics from your topic outline (p. 18). Headings are usually known to editors and publishers as *first order* (such as 'Methods', 'Results', 'Discussion'), *second order* ('Determination of X'), *third order* ('Preparation of reagent A [for determination of X]'), and so on. Keeping your hierarchy of headings straight will be easy if you write '1st', '2nd', etc. beside them in the margin of each draft. Make sure that a topic assigned to a given level of heading is truly equivalent in the hierarchy to others assigned to the same level. Keep the headings short.

The Introduction

Make the introduction brief, remembering that you are not writing a review article: two or three paragraphs are usually enough. Indicate the aim and scope of the paper. State your purpose in undertaking the work and—in appropriate disciplines—where it was done. Explain how your investigation moves forward from closely related, previous work on the same subject. Be concise but clear: aim to awaken interest rather than stifle it with fussy detail, and try to gain and keep the attention of readers who are not specialists in your field.

People who find writing difficult—that is, most of us—sometimes fill introductions with platitudinous general statements. There is nothing wrong with doing this if it clears your mind and helps you to start writing. But there is no good reason for publishing these generalities and you should remove them ruthlessly before you reach the final draft. A drastic piece of advice worth considering is: tear up the first two pages.

The 'Materials and Methods' Section

Unless the overall experimental design or theoretical approach is already obvious from the introduction, describe it in broad outline *before* you give details of the methods. State the premises and assumptions made in the design, and justify your choice of any methods (including statistical methods) to which there are reasonable alternatives.

Follow a logical order in describing the methods, and provide enough details for an experienced investigator to repeat the experiments, or at least to assess the reliability of the methods and therefore the results. Often, the best order is chronological: first procedures first, and so on. If you decide instead to describe different techniques in turn, be careful not to confuse the readers. They will be mystified if you describe how to assay a substance whose existence or relevance has not yet been mentioned.

In this section, change your mental picture of 'the readers' temporarily and regard them as colleagues with research experience similar to your own, so that you do not describe familiar items or procedures in tiresome detail.

Use your judgement about how much the readers need to be told. If you have used methods which are described elsewhere, it is pointless to give full details again; refer to a published text instead. Do, however, orientate the readers by stating the principle on which these methods are based, unless the methods are so well known that this would be naive. If you have used a previously published method in a modified form, describe the modification: 'a modified Kjeldahl determination' is uninformative.

In descriptive work—for example, in botany, biogeography or anthropology—the Methods section may contain itineraries or other chronological and geographical matter in some detail. For infrequent surveys especially it may be desirable to comment quite fully on earlier work and methods in the same region.

Plan to give the names and short addresses (town and county, state or province, etc.) of the manufacturers or suppliers of equipment or drugs with registered trade names. Use trade names for chemical compounds (e.g. pesticides) only if this is the most accurate way of identifying what you used, and quote the full chemical name on first mention of the trade name. In general, use systematic chemical names or the pharmacological names recognized in the country where your target journal is published (see the national pharmacopoeia). The World Health Organization periodically issues recommended or proposed international non-proprietary names (see also Marler 1973). State what tests for purity of chemicals were applied and say what the interval was between performing the tests and conducting the experiments. For enzymes, give the systematic name and the Enzyme Commission number at first mention of the enzyme; after that you may use the trivial name (Commission on Biochemical Nomenclature 1973).

Give the genus, species, race, strain, breed, cultivar or line of any experimental animals, plants or microorganisms used. Identify the authority for your system of nomenclature, if necessary. 'Rat' is not a precise term for a laboratory animal. State the age, sex, source, diet, dietary status, and condition of animals even if these seem only marginally relevant to your study. Frequency of handling, duration of light–dark cycle, and mixing of the sexes in the same room can affect the results. For many organisms, the name of the source or supplier is essential information.

In reports of experiments on human beings or higher animals discuss ethical considerations and safeguards explicitly in this section. Describe criteria for the selection of experimental and control subjects and the method of obtaining informed consent (from humans) and give the name of the ethical committee whose approval was obtained. These matters have often been taken for granted in the past, but nowadays they constitute a most important part of the 'Methods'.

If you used new or unusual equipment, describe it in this section.

The 'Results' Section

Make the results section comprehensible and coherent on its own. Even if you are planning to write a detailed discussion section later, do not merely describe here a series of experiments without any indication of their purpose, significance and relevance to your line of thought. Allow yourself to make connections! On the other hand, if you do not intend to discuss the results extensively, combine the results and discussion sections. In descriptive work, the results themselves are often the object of the investigation and need no discussion beyond a brief comment as you present them.

Present your results in a logical order, using only observations that are strictly pertinent to your argument. Report any negative results which could be important to other workers. Write your text in relation to the tables and illustrations you drafted earlier (pp. 10–15), and draw the reader's attention to the main points of your observations as interestingly as possible: instead of writing 'Results of Experiment A are given in Table 1' say 'Table 1 shows that the insects' behaviour is disturbed less than that of the rodents.' But do not repeat in the text the numerical evidence contained in tables and illustrations. Above all, do not tediously describe those tables and illustrations as though they were invisible.

In quantitative work, results are usually presented as values that have been derived from the actual measurements by a process called reduction of the data. It is your responsibility (*a*) to explain the method used for converting measurements into results, (*b*) to report results in a form as closely related to experimentally observed quantities as practicable and in such a manner that the degree of experimental variation can be assessed, and (*c*) to give estimates of the precision and accuracy of the results. This entails giving an example of a measurement together with its conversion into a result if data reduction was at all complex;

giving standard errors of the mean or standard deviations and the number of measurements on which means are based; describing statistical operations exactly, and your justification for using them; and making explicit your assumptions about the experiments as well as describing any auxiliary experiments done to support those assumptions.

Add diagrams wherever these will help to clarify the design of an experiment or a hypothesis emerging from the results.

At this stage, if you have the opportunity, you might present and discuss your methods and results at a departmental seminar or informal (unpublished) society meeting. The comments you hear will indicate whether you need to do more work before you draft the discussion section.

The 'Discussion' Section

A discussion is 'a disquisition in which a subject is treated from different sides' (Shorter Oxford English Dictionary). Write this section only after you have thought long and hard about your own and other people's findings. Assess the validity of your results, comment on their significance, and relate them to previous work. Do not simply repeat, in a different order, what you have already said or shown in the results section. Do not conceal negative results or discrepancies between your own work and that of others; try instead to explain them, or else admit your inability to do so. Criticize the scientific basis of other people's work when you feel it necessary to do so, but do not attack the authors personally. Be absolutely accurate when you describe or quote from other people's work (see p. 36).

Pull the threads of your argument together in logical form in the discussion. Do not reconsider every scrap of your work in minute detail. Refer to the tables and figures only as they become appropriate to the argument, and not necessarily in chronological order. Point out any results that suggest new lines of study. Try to explain the findings on the basis of one or two hypotheses rather than a multitude. If you want to speculate, make it clear where facts end and speculation begins; be concise, stay close to experimental observations, and restrict yourself to speculations that can be tested.

Do not be over-optimistic in your claims about the precision of your work, the generality of the conclusions, or the applicability of the results.

The Summary

A summary, if provided, is for people who have already read the whole paper; it should not be a re-worded abstract. Include a summary only if the journal specifically asks for one instead of or in addition to an abstract. State your main findings and conclusions, and refer only briefly to new hypotheses and future work. Unlike an abstract, a summary may include references to figures and tables in the paper itself. Occasionally, the journal may request a summary in a language other than English.

Acknowledgements

Acknowledge briefly any substantial help you received from grant-giving bodies or from individuals who supplied money, materials, technical assistance, or advice on the conduct of the work or preparation of the paper. Acknowledge the cooperation of departments or colleagues who provided specimens, referred patients, performed assays or provided other help not forming part of their routine obligations. Be sure that all those you thank agree to having their help recognized and that they approve the form in which you acknowledge it. Place the acknowledgements at the end of the paper, before the list of references, unless the journal prefers them elsewhere.

Appendixes

If you decided to include an appendix (see p. 10), place it here or after the references, according to the practice of the journal. Incorporate any references in it into the main list of references.

References

Collect details of references on cards or in any other suitable way while you are writing the draft, but do not allow this activity to interrupt the flow of your writing. As soon as you have finished the first draft, check that you have full details of all references.

For references to journal articles, 'full details' means:

- the names and initials of all authors;
- year of publication;
- title of article;
- unabbreviated title of journal (and the series, if the journal is issued in two or more parallel series);
- volume number;
- first and last page numbers (if illustrations are not inserted in the text but are placed elsewhere, list these too, so that they will be included if you later ask for a photocopy of the article).

Write down the issue number as well as the volume number for your own information, to make retrieval of recent issues easier in the library.

For references to books or chapters in books, note:

the names and initials of all authors;
year of publication;
title of book;
names and initials of editors, if any;
number of the edition if it is the second or later;
first and last page numbers of the chapter or section referred to (including illustrations, if necessary);
name of publisher (unless the book was published too long ago—say over 100 years—for this information to be relevant);
town of publication (if several towns are named, give the first one only).

For references to publications which use the Cyrillic alphabet, transliterate according to Table 1.

If you have had to rely on an abstract, a translation, or a citation in another article instead of the original paper or book, list the secondary source with the bibliographical information you are collecting (adding the words 'cited by . . .' or 'abstract in . . .').

Keep the reference cards, reprints and so on with the rest of your draft material until later, rather than spending time now making a reference list which will need to be altered for the final version. Chapter 5 will discuss the form that references should eventually take. Unpublished material from public archives should be properly identified in the text or in a footnote, not in the list of references. If the material belongs to a private collection, the necessary permission must be obtained.

We recommend (and some journals require) that you should refer to unpublished work in the text only, and not in the reference list. The reason for this recommendation is that readers ought not to be misled, however briefly, into thinking that you are supporting your statements with references to published work which they can examine for themselves. Obtain permission from the cited person(s) to refer to such work, for the reasons already given (p. 16).

Although doctoral and other theses may be included in the reference list, you should refer to technical reports or similar documents of limited circulation—whether issued by academic institutions, government agencies or commercial companies—only if you have obtained permission to do so. If the documents you are allowed to cite are available on request, record full details of how to obtain them. If they are not available on request, cite them only in the text, as unpublished work.

TABLE 1

Transliteration of Slavic Cyrillic Alphabets*

	Transliteration from					
printed	Russian	Ukrainian	Byelorussian	Serbian	Bulgarian	Macedonian
а А	a	a	a	a	a	a
б Б	b	b	b	b	b	b
в В	v	v	v	v	v	v
г Г	g	g	g	g	g	g
д Д	d	d	d	d	d	d
Ђ ђ				dj		
ѓ Ѓ						g
е(ё) Е(Ё)	e	e	e	e	e	e
є Є		je				
ж Ж	zh	zh	zh	zh	zh	zh
з З	z	z	z	z	z	z
ѕ Ѕ						dz
и И	i	i	i	i	i	i
і І		yi	yi			
ї Ї		yi				
ј Ј				j		j
й Й	j	j	j		j	
к К	k	k	k	k	k	k
л Л	l	l	l	l	l	l
љ Љ				lj		lj
м М	m	m	m	m	m	m
н Н	n	n	n	n	n	n
њ Њ				nj		nj

	Transliteration from					
printed	Russian	Ukrainian	Byelorussian	Serbian	Bulgarian	Macedonian
о О	o	o	o	o	o	o
п П	p	p	p	p	p	p
р Р	r	r	r	r	r	r
с С	s	s	s	s	s	s
т Т	t	t	t	t	t	t
ћ Ћ				cj		
ќ Ќ						k
у У	u	u	u	u	u	u
ў Ў			w			
ф Ф	f	f	f	f	f	f
х Х	kh	kh	kh	kh	kh	kh
ц Ц	ts	ts	ts	ts	ts	ts
ч Ч	ch	ch	ch	ch	ch	ch
џ Џ				dzh		dzh
ш Ш	sh	sh	sh	sh	sh	sh
ш Ш	shch	shch	.		shch	
ъ Ъ	″	″	″		y	
ы Ы	y		y			
ь Ь	′	′	′		′	
э Э	eh		eh			
ю Ю	yu	yu	yu		yu	
я Я	ya	ya	ya		ya	

* Based on ISO Recommendation R9–1968. Reprinted, with permission, from the International List of Periodical Title Word Abbreviations.

When you have written the first draft, put it away for a day or two. Then re-read it quickly and not too critically, correcting only the obvious mistakes as you go through it. Rewrite or retype any pages that have become illegible. Number the pages and date them or use differently coloured paper to show which version is the most recent. File the superseded pages in case you have accidentally omitted something on them which you later want to restore. Before you embark on the next stage—revising the draft—put the manuscript away and forget about it for a week or two.

Chapter 4

Revising

A good way to revise the first draft is in two stages: structural and stylistic. It is a waste of time trying to improve stylistic details before you are sure that the sections, paragraphs and sentences are in the right order, that all the essential points have been included and any superfluous ones removed, and that the argument runs logically from hypothesis to conclusions. Start with the structure, therefore, and examine everything you have so far prepared for logical order, accuracy, consistency and truth. For guidance on scientific logic and argument, we strongly recommend Trelease (1969).

Examine the Sequence

If your outlines were satisfactory, all the statements in your draft probably contribute something, and no points will have been forgotten. But in getting everything quickly onto paper you may have strayed away from the main line of your argument, introduced unnecessary material, or left out essential evidence. In addition, you may now realize that some points need to be discussed or explained earlier in the article than your outlines originally suggested. Check on these matters, especially on the need to move some passages to an earlier place in the text to give increased clarity. Now examine the headings (*a*) to see whether they relate properly to one another and to the text they describe, and (*b*) to see whether any should be deleted. Make sure that each heading is appropriately ranked and clearly identified as first, second or third order.

As you read the typescript, note also the length of the paragraphs and the distribution of ideas among them. In principle, each paragraph should deal with a single topic or message and so be a unit of thought. However, readers find solid blocks of print tiring and you should give them a rest by keeping most paragraphs to not more than 125 words, or half a typewritten page. If many paragraphs occupy more than a page, look for places to break them up. But if most of them are only a few lines long, you may be making the mistake of letting each new sentence form a paragraph: group the sentences into paragraphs now. Thinking about paragraphs and their length like this gives you another way of examining the structure of the paper for logical flow.

Check the Illustrations and Tables and Prepare them in Final Form

Re-examine the illustrations and tables in the light of the argument and structure of paper as it is now. Are any of them redundant or less relevant than you thought at first? If so, remove them. Can any remaining illustrations or tables be combined for greater effect? Can any be simplified? Make any necessary changes—and the appropriate alterations to titles, legends and footnotes—before you prepare the final versions. This reconsideration of your results will help to give you a new view of the text.

Tables

Next type the tables—or have them typed—in their final form, in the way described on p. 13 and pp. 64–65. Compare the titles and explanatory notes of all the tables throughout the series, and check that symbols are consistently used and are linked to the right explanatory notes. Check that all abbreviations have been explained. Make sure that you refer to the same substances by the same names in every table. Do not arbitrarily change the word order in titles of similar tables.

Illustrations

Prepare the illustrations, or have them professionally drawn or finished, in the way described in the next few paragraphs and in accordance with the journal's Instructions to Authors. Show the relevant parts of the Instructions and our recommendations here and in chapter 2 (pp. 13–15) to the technical artist or photographer, as necessary. Figs. 1–8 provide examples of good and bad ways of presenting illustrations.

Illustrations are usually reduced during the printing process to a size that conveniently fits a column or page of the journal. Calculate what size your drawings and photographs will be when thus reduced, and make the originals not more than three times as large. Ensure that the proportions and the lettering will come out in the right size after reduction (see p. 31). Ease of handling is secured by making the dimensions not larger than the typescript (A4 paper = 210 × 297 mm). A reducing glass—a biconcave lens—will help you to judge what the final result will be.

Sometimes circumstances demand a larger original, up to five or six times the desired final size. The production of such an original needs careful calculation, not only of the size of elements within it, but also of the thickness and spacing of lines: details that are too fine do not register, while those that are drawn too heavily may come out solid black.

To avoid having large photographs reduced and essential details obscured, select the parts you want the reader to see and either cut off or mask the rest (Figs. 7, 8). Even when you do not expect illustrations to be reduced, add scale bars to all maps, micrographs etc. so that absolute sizes are clearly recorded and any change in size during printing is automatically compensated for.

If several of your line drawings represent similar material, draw them all to the same scale (Figs. 1, 2). According to the journal's preference, use good-quality stiff paper, white card, tracing paper or cloth, or graph paper with faint blue lines (other colours are liable to be reproduced in the printed figure). Do not use celluloid. Use black Indian ink only, and stir it: unstirred ink may be grey and will reproduce patchily, if at all. Avoid the kind of graph paper which has printed lines that repel the ink. Draw lines neither too thick nor too thin; for broken lines draw solid lines first, then break them with white ink, or use Letraset or other preprinted lines. Do not draw a rectangle around a drawing (see Fig. 1). Leave a reasonable space—not too much—between the *x* axis of a graph and the word or words describing it, and letter the left and right axes as shown in Fig. 4 (cf. Fig. 3). Note that it is not necessary to number every division on the axes (Figs. 1, 2).

The journal's instructions should tell you whether to put lettering and other symbols on the original illustration or whether to pencil them on a transparent overlay (or write them in blue crayon on a line drawing) as a guide for the printer. Make letters and symbols large enough to be *at least* 1.5 mm high after reduction. The lettering on axes of graphs should be about 2–3 mm high after reduction, while any numbers or letters identifying an illustration may be as much as 4 mm high after reduction (Fig. 2; cf. Fig. 1). Avoid heavy black ('bold') lettering, as this will look unpleasantly dark when printed (Figs. 3, 5). Do not make cross-hatching too fine, especially on large drawings, or it will come out solid black (see Fig. 1).

Symbols and abbreviations must be those approved by the journal. Remember to place a zero in front of a decimal point (0.1, not .1), to change μ (for microns) to μm (for micrometres) and to convert Å (ångströms) into nm (nanometres) if the journal uses SI units, as it should. Use the journal's pre-

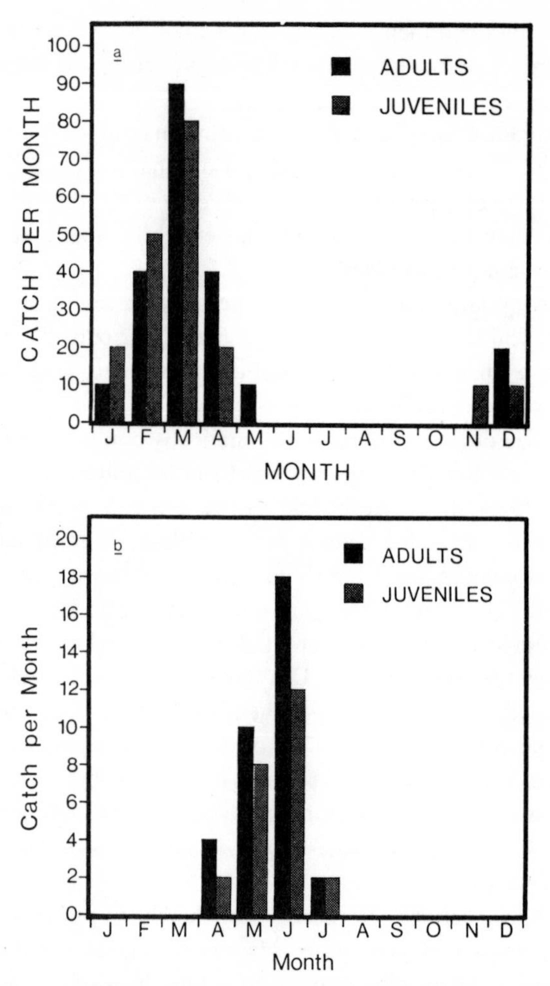

Fig. 1. Common faults of draughtsmanship. 1, scales different in *a* and *b*; 2, insufficient contrast between columns for adults and juveniles; 3, inconsistencies in type size and use of upper and lower case; 4, symbols *a* and *b* far too small; 5, excessive numbering on *y* axis; 6, heavy black frame.
(*Note:* Legends explaining the *meaning* of Figs. 1–8 have been omitted for the sake of clarity.)

ferred symbols (usually ○, △, □, ●, ▲, ■) to indicate experimental points on line drawings, and differentiate curves by different point symbols joined by the same type of line (○——○, △——△) (Fig. 4) *or* by the same symbol joined by different types of lines (●——●), (●- - -●) (*not* as in Fig. 3). Place the key in the legend, normally, but see what the journal recommends. If you have to include non-standard symbols which the printer may be unable to reproduce, place the key in the body of the figure (Figs. 2, 4).

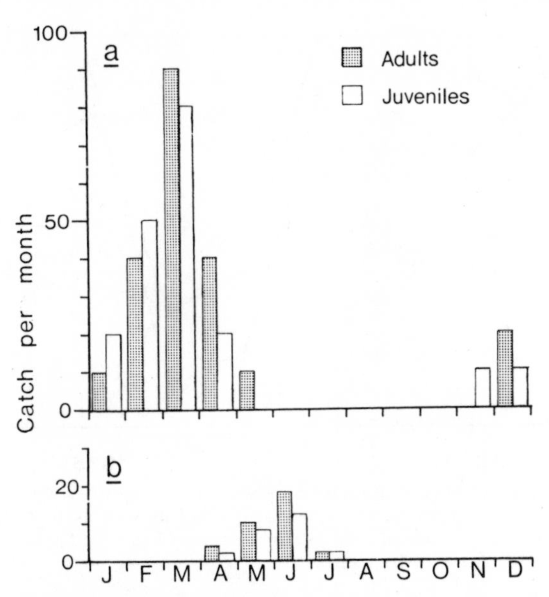

Fig. 2. Clear presentation of the data in Fig. 1. 1, histograms on the same scale; 2, economy of line and lettering; 3, contrasting columns for adults and juveniles; 4, correct use of type sizes to highlight the important features.

On axes of graphs and elsewhere, use lower-case, not capital, letters, for whole words (Figs. 1, 2). If you make a mistake in drawing or lettering, put white paper or ink over the mistake and start again; if you do this carefully it will not show up in the printed illustration. Use a stencil set or preprinted letters such as Letraset, if available. Never draw letters freehand—in particular, do not change 'u' to 'μ' by hand—unless you can do this to a professional standard (Fig. 8; cf. Fig. 7). If you run out of a preprinted letter, invest in another printed sheet rather than draw the letter in (see Fig. 5)—and be sure to use the same style of letters throughout (cf. Figs. 3 and 4). Burnish and spray preprinted letters so that they remain fixed on the illustration.

On photographs with a dark background, use black letters or symbols on white labels, and stick these on securely (Fig. 8).

Photographs which you want grouped as a single plate should be mounted on card (do not mount single photographs). Use stiff white card and ensure that all the corners of the prints are right angles and that the outside edges form a square or rectangle (Fig. 8). Arrange the photographs on the card so that the whole plate is in proportion to the journal's page size; match the plate to the width of the page if it cannot be matched to both width and length. Do not allow the card to show between the photographs and do not paste dividing strips between them; the printer will mark any divisions that are necessary. The

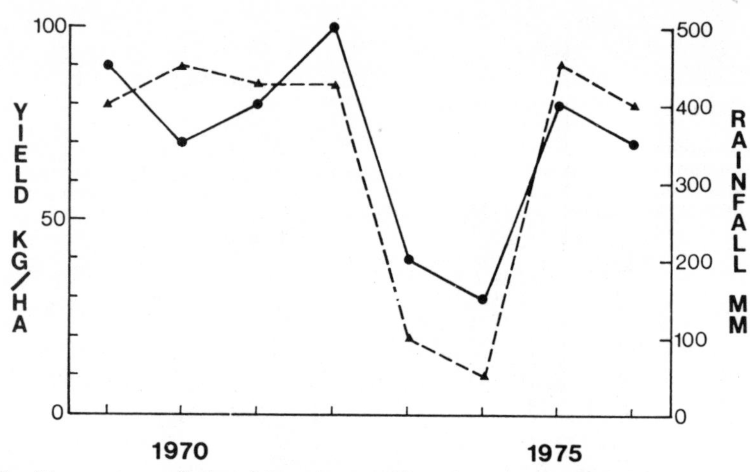

Fig. 3. More common faults of draughtsmanship. 1, no key to symbols; 2, lines joined to the points; 3, the two graphs distinguished by symbol as well as type of line; 4, vertical lettering difficult to read; 5, use of bold lettering detracts from the appearance of the graph; 6, numbered years too far below *x* axis.

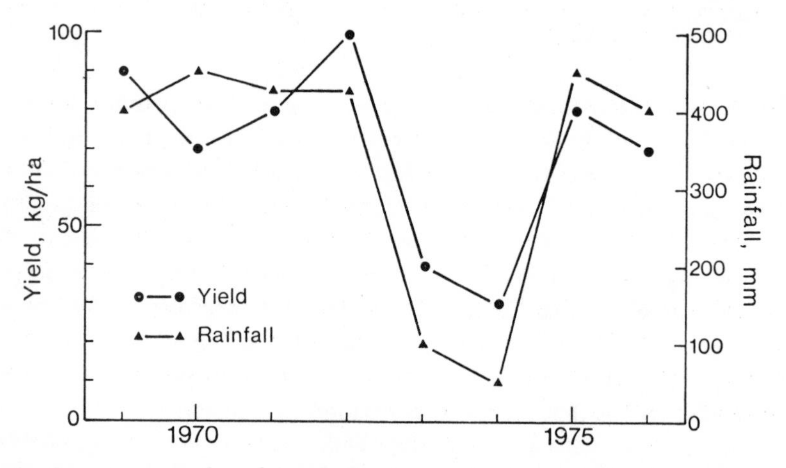

Fig. 4. Correct presentation of data in Fig. 3. 1, a key included; 2, points made to stand out by breaking the lines; 3, graphs distinguished by symbol only; 4, correct positioning and arrangement of lettering.

journal usually specifies whether you should supply glossy or matt prints, but if no instruction is given, submit glossy ones.

Submit the original artwork for line drawings, in preference to photographs, though good, well-focused photographs lettered exactly as required by the journal are sometimes acceptable (see p. 13).

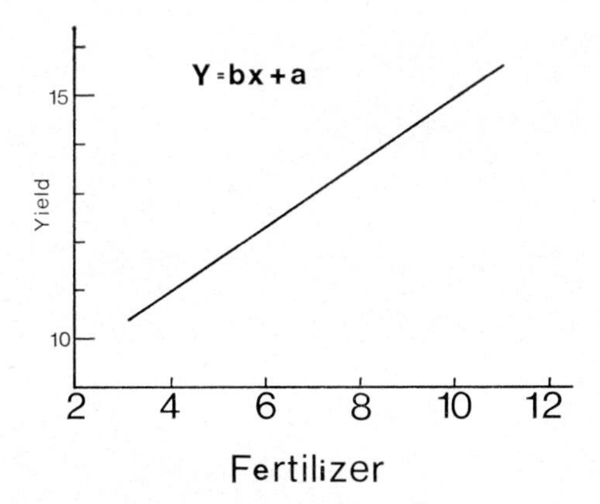

Fig. 5. An uninformative regression line, with faults of draughtsmanship. 1, no units on x and y axes; 2, no data on derivation or accuracy of regression; 3, lettering on y axis too small, on x axis too large; 4, bold type for equation placing emphasis in the wrong place; 5, mixed type sizes in the word 'Fertilizer'.

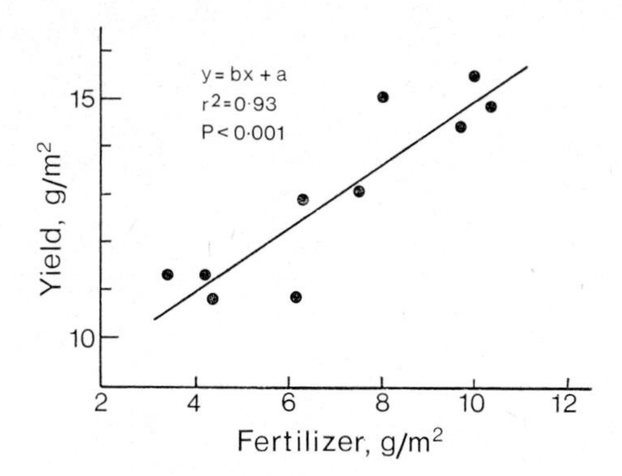

Fig. 6. Informative and clear presentation of Fig. 5. 1, units on x and y axes; 2, individual points and regression data given; 3, lettering of appropriate size and weight.

When all the illustrations are ready, check the legends against these final versions. If the illustrations have been renumbered since you drafted the legends, make sure that the legends have been renumbered too, and check that all the symbols and abbreviations have been explained. Compare the wording of the legends and eliminate unnecessary words and inconsistent statements.

Re-Read the Papers Cited

Next read right through all the articles you have cited in your draft. Memory plays tricks, and relying on memory alone will put you in danger of misquoting or misrepresenting work you read perhaps years ago—even your own! Do not

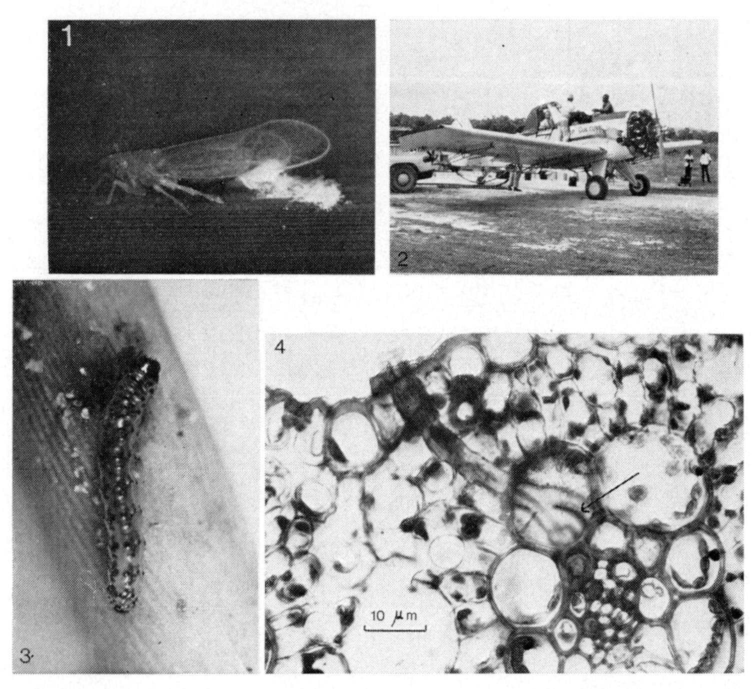

Fig. 7. An incorrectly prepared plate. 1, the outer edges not a rectangle (this doubles the cost of block-making); 2, uneven spacing between photographs; 3, numbering on photographs inconsistent in terms of colour, size and position; 4, hand-drawn symbols (arrow and μ) suggest carelessness.

neglect the methods sections: you may be surprised to find how much your current methods differ from those you thought you were imitating exactly. Correct your text accordingly.

Make sure that the work you refer to is indeed relevant to the points you make. Some authors used to take pride in providing a long list of references, but knowledgeable readers are unfavourably impressed by the extravagant citation of material only marginally related to what they are reading. Readers will also think better of you if you cite other people's work more often than your own.

If you quote a passage, do so in context to avoid distorting its meaning. Keep to both the spirit and the letter of the original, copying any mistakes exactly as printed. You may insert '[sic]' after any word you think was misspelt or misused in the original. Place anything that you add to the quoted passage between square brackets. Use three spaced stops (. . .) to show where words have been left out. If the passage is not in English, quote it in the original language and add a translation.

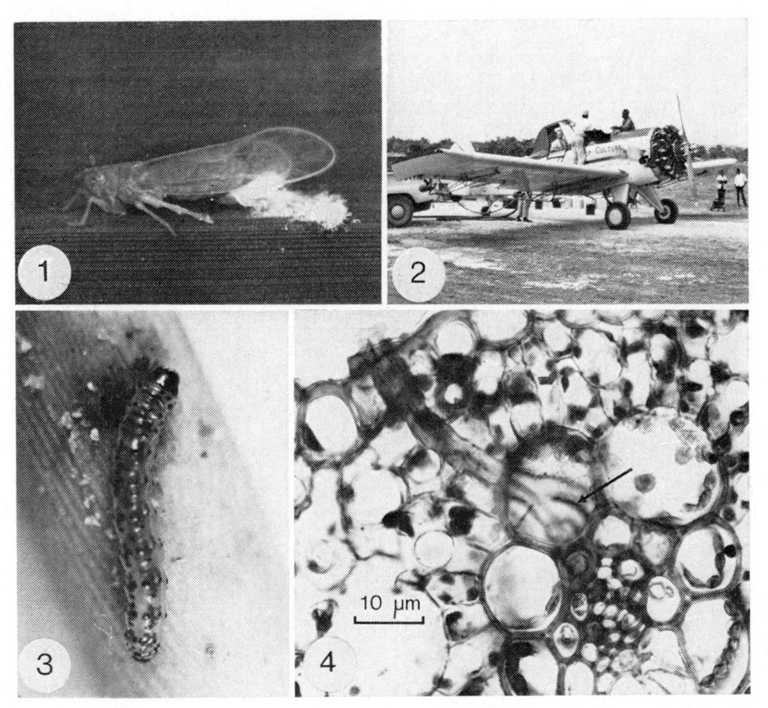

Fig. 8. The same photographs as in Fig. 7 correctly laid out. 1, photographs trimmed to make a rectangle and to give even spacing between them (note, there is no loss of information); 2, numbering on photographs clear and consistent; 3, preprinted lines and symbols give better finish to the plate.

Get permission from the copyright-holder for quotations of more than 50 words or so (see p. 15).

Check for Consistency

Now check whether you have been accurate in other matters. Have you spelt the same word in the same (correct) way throughout?* Have you written numbers, symbols and abbreviations in the form approved by the journal?

* Many English words have British and American spellings. Neither is 'more correct' than the other, but you should be consistent. A good dictionary (see p. 43) will show both forms of spelling and tell you which is which. Some journals prefer British, others American spelling, but many will accept either kind of spelling as long as it is used consistently throughout a paper.

Have you observed the rest of the journal's conventions? Check your text against the Instructions to Authors. Inconsistency in these matters may not worry the editor or referees as much as inaccuracy in citations—but neither is it likely to prejudice them in your favour.

Reduce the Number of Abbreviations and Footnotes

Many otherwise well-written scientific papers are marred by being overloaded with abbreviations ('We added DTNB to the TCA-precipitated RNA before determining the BUN'). Use abbreviations sparingly, as substitutes only for unwieldy terms that have to appear very often (say more than ten times) in the paper. Define them either at their first appearance, or all together in a footnote at the beginning of the paper, or both, according to the journal's requirements. Try to replace frequently repeated abbreviations with pronouns or other substituting words ('it', 'the enzyme', and so on), but do not use this device if it leads to ambiguity. Make sure that the abbreviations used are acceptable by checking them in Appendixes 3 and 4 or in the lists provided by some journals and scientific societies. Remember that chemical *symbols*, being internationally understandable and unambiguous, are nearly always acceptable, so that—for example—CCl_3COOH is preferable to TCA (which is often used to denote trichloroacetic acid but can also denote tricarbonic acid).

Avoid using footnotes. They are justified only rarely, when you need to give subsidiary information that would otherwise seriously interrupt your argument.

Type the First Draft

Now type the manuscript, or have it typed, so that you have a clean copy to work on in revising the style. Type it on good paper, with wide spacing and wide margins all round, and make at least one extra copy which can be filed for safe keeping while you work on the top copy.

Revise the Style

Editors do not, or should not, expect people brought up speaking other languages to write perfect English, but they do expect all authors to express themselves as clearly as possible, avoiding clumsy phrases and major grammatical errors. In the rest of this chapter, therefore, we try to show you how to escape the main pitfalls of English grammar—the problems that give trouble even to many whose mother tongue is English. Our treatment of these problems is necessarily inadequate, as this guide is not intended to be a textbook of

English grammar, and we suggest that you consult a good modern textbook if you need further guidance. (Hornby's *A Guide to Patterns and Usage in English*, 1954, prepared especially for those writing English as a foreign language, is both useful and cheap. Tichy's *Effective Writing*, 1966, provides excellent advice and good examples of well-written passages, while Strunk & White's *The Elements of Style*, 1972, is superb on the art of writing clear, simple English.)

Throughout the text, transform laboratory, hospital or workshop jargon into universally understood phraseology. It is the author's responsibility to use up-to-date terminology and nomenclature—and to use them correctly—when referring to units of measurement, substances, plants, animals, tissue components, etc., especially if international conventions exist on those matters.

Problems with verbs

(*a*) Non-agreement of subject and verb. 'The enzymic activity of the pollinaceous extracts from mature plants were higher than expected' should read '*was* higher' since the subject of the sentence is 'activity', not 'extracts' or 'plants'. Deal with this fault by examining every verb in the manuscript and asking yourself what its subject is.

(*b*) Incorrect omission of 'was' and 'were' (auxiliary verbs) in a series of passive verbs. 'The sample was weighed and several fractions taken for examination' is wrong because 'fractions' as the subject of the second verb needs the plural auxiliary verb 'were', not the implied singular verb 'was'. 'The samples were weighed and the heaviest fractions taken for analysis' is correct, but be careful about omitting an auxiliary verb even when both subjects in such a sentence are plural; this sometimes makes it sound as though the second verb is intransitive when it is not: 'The mice were decapitated and their diaphragms stretched' leaves the reader wondering whether the diaphragms stretched of their own accord or were stretched.

(*c*) Dangling, hanging or unattached participles. When you are writing in the third person, avoid participles as much as possible or you will find that you have attributed human actions to inanimate objects, animals or plants. 'Examining [present participle] the results, the conclusions are obvious' means, if taken at its grammatical face value, that the conclusions are examining the results. Because the participial phrase 'examining the results' is grammatically unattached it appears, absurdly, to belong to the subject ('the conclusions') of the main clause. Correct this fault by substituting an active verb for the participle and naming the subject: 'When we examined the results, the conclusions were obvious'. Check all words ending in 'ing' in the text and see whether they need a subject; if so, see whether that subject is correctly and unambiguously identified. The word 'following', when used as a preposition ('following incubation') often sounds like a dangling participle and can usually be replaced advantageously by 'after', which is unambiguous. 'Following' is correctly used as an adjective, for example in 'the following recommendations'.

Unfortunately, gerunds (verb-nouns) have exactly the same form as the present participle in English. Because a gerund is a noun it can, unlike the present participle, be the subject of a sentence: 'Acquiring [gerund] this instrument proved difficult' is correct, but 'Before acquiring [participle] the instrument, the rats were housed in constant-temperature chambers' is not—it implies that the rats acquired the instrument.

(*d*) Dangling infinitives are just as dangerous as participles, because the unstated, understood subject of an infinitive may not be the subject of the next clause: 'To apply this form of treatment, the patient had to be admitted to hospital' is wrong because 'patient' is not the subject of 'apply'. Correct this fault by substituting the true subject: 'To apply this form of treatment, we had to admit the patient to hospital'. Note that adding 'In order' at the beginning of the original sentence would not correct the error, as the infinitive would still not be attached to a subject. Note also that the 'corrected' sentence 'To apply this form of treatment, admission to hospital is necessary' still contains a dangling infinitive ('admission' is not the subject of 'to apply') and is just as bad as the original sentence.

(*e*) Excessive use of the passive voice. It is absurd, and mentally lazy, to use the passive voice in every sentence. For 'the membrane is crossed by the protein' write 'the protein crosses the membrane': it is shorter, clearer and more correct —except in the rare instance when you want to emphasize what kinds of things cross the membrane. Write 'Fig. 1 shows', not 'As is shown by Fig. 1'. In the Introduction or Discussion sections of papers write 'we believe' rather than 'it is thought' if you are referring to your own thoughts. (This type of circumlocution has been called the 'passive of modesty'.) If you think you must use 'it is generally thought', try to identify *by whom* it is generally thought: you will probably decide to abandon the phrase. The right place for a verb in the passive is in a sentence like 'The cows were milked twice a day' when there is no need to identify the milkmaid, or in a sentence like 'The cows should be milked twice a day' when there is no need to direct any particular person to do that job. The passive voice is appropriate in many sentences of a scientific paper, but you should change to the active voice whenever possible, if only to keep the reader awake by varying the rhythm! Many journals now encourage authors to use the first person, active voice ('I tested the hypothesis by means of the following experiments') which for decades was shunned as indicating too subjective an approach. Unfortunately, that restriction often led to a semblance of objectivity which tended to bring science into disrepute.

Problems with pronouns

The words 'it', 'this', 'that' and 'which' can give trouble if they do not refer unambiguously to an antecedent noun. Find every such pronoun in your text and ensure that the reader will immediately understand what it refers to. It is far better to repeat a noun than to risk being misunderstood.

Problems with nouns

(*a*) Over-use of abstract nouns instead of verbs. Some scientists have a mania for using abstract nouns. This leads to a heavy and colourless style which is often mistakenly regarded as a model of objectivity and detachment. Good English stylists detest this habit. If the only English you read is that in scientific journals, be especially careful: your style is undoubtedly so contaminated that you are probably writing sentences like this: 'The addition of x and y to the medium permitted shaking of the solution without the formation of ice crystals and the precipitation of z.' The words in this sentence ending in '-tion' —*addition*, *formation* and *precipitation*—are abstract nouns derived from the verbs *add*, *form* and *precipitate*. Note that abstract nouns are almost always longer than their corresponding verbs; both their length and their abstract quality contribute to the ponderous effect for which you are probably striving in imitation of your British and American models. Stop striving and write 'After adding x and y to the medium we were able to shake the solution without causing ice crystals to form and without precipitating z.' Any narrative written in abstract nouns is clogged with words like *of* and *the*. Many of them disappear when the abstract nouns are replaced by the corresponding verbs. Learn to suspect that inside every abstract noun there is an active verb struggling to be let out. The distress signals in your text will include weak and colourless past participles such as *occurred*, *effected*, *brought about*, *achieved* and *produced* (see Appendix 5, p. 94). Having detected the fault, you can improve sentences like 'Separation of A from B was effected (brought about, achieved, performed, obtained, carried out, etc.)' by writing 'A was separated from B' or 'we separated A from B' instead.

(*b*) Excessive use of nouns as adjectives (modifiers). English is flexible in allowing nouns to be used to modify other nouns: 'protein iron', for instance, is a short way of referring to the iron bound to or contained in proteins. The trouble begins when, perhaps in a laudable effort to be brief, the writer puts together a string of nouns, each of which modifies one of the others. Phrases like 'adult sheep muscle protein iron' impede understanding and may even defy it. Break up these clusters by inserting verbs and prepositions: 'protein iron *found in* the muscle tissue of adult sheep'. Look with suspicion at strings of three or more nouns—or for that matter at any noun with two or more modifiers—and separate the cluster to give increased precision as well as clarity.

Problems of verbosity and pomposity

Some people will try to persuade you that formal English, suitable for scientific papers, must use polysyllabic, Latin-derived words and an 'elevated' tone if it is to be effective. Take no notice of this advice: it is wrong. The most effective writing, in a scientific journal or anywhere else, is simple, clear, precise and vigorous. If you want to write effective prose, search for the simplest, most direct way to express your thoughts.

Be brave and say 'It is most often found in the heart', not 'The most frequent among its localizations is the cardiac one' (and remember to see that 'it' refers unambiguously to a noun or phrase in the preceding sentence). Beware of hedging over uncertainties or suppositions: 'It may seem reasonable to suggest that necrotic effects may possibly be due to involvement of some toxin-like substances' contains eight degrees of uncertainty and only means 'Necrosis may be due to toxins'. And do not suddenly hedge or hesitate after a positive phrase: what does 'This strongly suggests that female odour probably plays no part in . . .' mean?

English favours shorter sentences than many other languages. Shorten or simplify every sentence, provided that this can be done without sacrificing subtlety and exactness of meaning. Try to keep sentences to less than 40 words, but vary their length. It is more important to have a single idea in a sentence than to strive for constant length; besides, a series of 20-word sentences is almost as tedious and distracting as a succession of 80-word ones. Always place different ideas in separate, clearly constructed sentences.

Choose common, short words in preference to long or archaic ones ('end', not 'terminate'; 'on', not 'pertaining to'; 'before', not 'prior to'). Cut out all long-winded introductory phrases like 'it is of interest to note that' and 'it will be seen upon examination of Table 2 that'. Such a throat-clearing exercise as 'Studies in Smith's laboratory some years ago by Schettler & Panotti (1966) showed that' should be thrown out in favour of 'Schettler & Panotti (1966) showed that' unless there is a good reason for mentioning Smith's laboratory. Remove any unnecessary adjectives and adverbs, giving special attention to vague qualifiers such as 'very', 'quite', 'rather', 'fairly', 'relatively', 'comparatively', 'several' and 'much'. 'Relatively' is particularly objectionable unless you really are *relating* one number or quantity to another.

Problems of imprecision

If you always choose elaborate words instead of simple ones you will soon be trapped into using words *wrongly*. English is so rich and complex that vigilance and self-discipline are needed for success in writing it correctly. The utmost accuracy in writing, as well as in experimenting or observing, should be every scientist's aim.

Make sure, therefore, that you know the exact meaning of every word in your text. Since this is hard, time-consuming work, the fewer words there are the better for you. You need good dictionaries, of course, including both a dictionary from your own language into English and a large English-language dictionary. It is often worthwhile looking up an important word in a dictionary from your own language into English, then checking the English words in a large English dictionary, and finally looking up the English word you select in a dictionary which translates back into your own language. The result can be astonishing.

Recent or recently revised dictionaries contain proportionally more technical terms than earlier dictionaries. This is why, of the English-language dictionaries, *Webster's Third New International Dictionary* is more useful in scientific writing than the *Shorter Oxford English Dictionary*, although British journals may use the Oxford dictionary as their authority for spelling. Another American dictionary, the *Random House Dictionary* (1966), is more recent than either Webster or the Shorter Oxford and is excellent (it gives British spelling as a permitted alternative). In addition, you can study English usage in *The Complete Plain Words* (Gowers 1973) and *A Dictionary of Modern English Usage* (Fowler 1965) if you want to be a really good writer, or in the *CBE Style Manual* (Council of Biology Editors 1972) or in *Scientific Writing for Graduate Students* (Woodford 1968) if you aim to write good plain scientific prose.

Even quite simple words may be used wrongly by people who do not bother to discover their exact meaning. If you want to point out that one value is lower than another, do not say it is 'decreased relative' to the second number: this is both cumbersome and wrong. 'Decrease' implies a change with time or after some event or manipulation, whereas the two values may have been constant for centuries. Mistakes like this are dangerous; in no time you will find yourself wondering whether the first number has in fact decreased. And your reader may be quite sure that it has. Again, when you write that a biochemical reaction proceeds by 'alternate' pathways, that literally means that it proceeds first by one pathway, then by another, then by the first, and so on; the word you probably want is 'alternative'. Only a careful study of the dictionary will enable you to distinguish 'affect' and 'effect' reliably. Do not imagine that 'content' and 'concentration' have the same meaning, or substitute 'level' for either or both of these words. The word 'parameter' is often misused (see Appendix 5). Have the courage to pin down your meaning precisely, without turning to escape-phrases like 'nature of the system', 'in this problem area', 'character of the situation', 'at the level of', or 'in the content field of this process'.

Because certain words and phrases are constantly misused even in reputable international journals, you may be unaware that you have acquired incorrect usages. Read through Appendix 5 now, and use your dictionary to establish the correct meaning of words you are surprised to find in the left-hand column.

Woodford (1968), in providing guidelines for solving problems of style, suggested the following principles:

Be simple and concise
Make sure of the meaning of every word
Use verbs instead of abstract nouns
Break up noun clusters and stacked modifiers
(that is, strings of adjectives and nouns, with
no clues as to what modifies what)

These principles summarize what we have just said. With their aid, you should be able to escape most of the stylistic weaknesses commonly found in scientific writing. If you can keep everything straightforward and concise, editors and referees—to say nothing of readers of the journal—will be grateful. Clarity and the orderly arrangement of ideas are far more important than perfect grammatical form—which is something the editor or someone else can help you to achieve if you have indicated clearly enough what you are trying to say.

Finally, do not try to carry over principles of stylistic elegance from your own language into English. Do not, for example, studiously avoid repeating a word in a sentence. Good English stylists prefer repetition to 'elegant variation' (Fowler 1965) because the real meaning can easily be distorted by a badly chosen synonym.

Problems of punctuation

Problems of punctuation are many but—in a way—trivial. If you write short, simple sentences you can avoid most pitfalls. Punctuate according to whatever rules of English usage you already know, always aiming to make your meaning clear and unambiguous to the editor, who can then improve the punctuation if necessary. The section on 'stops' in Fowler (1965), is helpful. Carey's book *Mind the Stop* (1971) provides more detailed discussion if you want it.

Note that the subject of a sentence must never be separated from its verb by a single comma, although commas must be used for a parenthetical clause such as 'The specimens, each of which was cruciform, were all found in the same place'.

A common cause of ambiguity is failure to punctuate an adjectival clause correctly: 'The barnacles which weighed over 10 g were . . .' (no commas) contains a *defining* clause implying that there were other barnacles which did not weigh that much; 'The barnacles, which weighed over 10 g, were . . .' (two commas) is a non-defining clause meaning that all the barnacles weighed over 10 g. It is preferable to use 'that' and no commas for a defining clause, 'which' and commas for a commenting or non-defining clause—read Fowler (1965, pp. 625–627) or Gowers (1973, pp. 142–143). If you stop every time you write 'which' and ask yourself whether it is the correct word, you will soon prevent ambiguity from this source.

Hyphens often give trouble. Use them to clarify meaning: 'The bee-filled hive was moved' is correct and clear when written with a hyphen; 'The bee filled hive was moved' is wrong and confusing.

Get Advice from Co-Authors and Friends

After you have revised the paper structurally and stylistically to the best of your ability, have the revised version (the *second draft*) typed again and show it (*a*) to your co-authors, if any; (*b*) if possible to sympathetic but critical colleagues in the same or similar fields—though not yet to the expert whose opinion you especially value and whose comments you are going to ask for eventually (chapter 5); and (*c*) if you can, to a friend in a different academic discipline. Give everyone complete typescripts, including copies of the illustrations, tables and collected references, since these are an integral part of the paper. Ask all these people to give you their comments in writing if possible; this usually saves everybody's time in the end, and keeps tempers cool. While the copies of the manuscript are being critically considered, relax and think about something else.

Chapter 5

Refining

When your co-authors and friends have read the second draft, think about their suggestions carefully and objectively. Read the draft critically yourself, and make any changes that you decide are needed. Before giving the article to a senior colleague or the head of your department for formal and detailed criticism, polish it into its near-final state. You will have to rewrite the title and abstract, list the index (or key) terms, check every part of the paper to ensure that it complies with the journal's Instructions to Authors, and construct the reference list in correct form.

Rewriting the Title and Abstract

You wrote the *working* title and abstract for yourself, as guides defining and limiting what you were going to write about (p. 8). Now write a title and abstract for publication, bearing in mind that the purpose of the published title is to bring the paper to the attention of the people who ought to see it, and that the purpose of the published abstract is to tell them succinctly what it is about and why it will interest them.

The final title

Make the title concise, accurate and informative. The object is to give the reader as much specific and intelligible information as you can in as few words as possible. 'Pollen morphology of *Saxifraga* hybrid *S. nathorstii* and its parents' has 67 characters (including spaces) and contains as much information as 'Notes on the pollen morphology of *Saxifraga* × *nathorstii* and its putative parents *S. aizoides* and *S. oppositifolia* (Saxifragaceae)', which has 130 characters. The journal's Instructions to Authors may specify a permitted or preferred length, although this will obviously vary with subject matter. If a title can be kept below 12 words and 100 characters (including spaces), so much the better. Readers find titles longer than this difficult to grasp when they are scanning a large number of them.

Many information-retrieval systems used by research workers now depend heavily on indexing by means of automated, computerized selection of words from the author's title. The trivial words ('of', 'the', 'on') are ignored; the non-trivial words in the title, and only these, are used for the indexing. In constructing your title, try to choose the words you would think of if you were looking for the paper in a subject index. 'Pollen' and '*Saxifraga*' are the two essential search words in the title quoted in the previous paragraph. A title must be neither too general nor too vague. 'Amino-acid composition' will not do, nor will 'Proteins in plants'. A title such as 'Amino-acid composition of proteins of the geranium leaf: changes with age and browning' (13 words and 86 characters) tells the reader much more clearly what to expect. If possible, begin with a significant word or phrase, specify the experimental material, indicate what has been measured, and hint at the significance of the work.

In your efforts to be informative, do not inflate the title into a short abstract—if you keep it to less than 100 characters you will avoid that danger. Keep the title short neither by using abbreviations (which are highly undesirable in titles) nor by accumulating modifiers until all sense is obscured (pp. 41–42), but by cutting out meaningless phrases such as 'Observations on . . .', 'A study of . . .', 'An approach to . . .', and similar expressions.

Give each paper you write a different title: that is, do not publish a series of numbered papers all with the same general title unless this is the only way to deal with a particular set of investigations. If it is important to link papers to previous papers in a series, you can do this by means of a footnote or by giving an appropriate reference in the first paragraph of the introduction.

Provide a short title—usually 45 to 60 characters—if the journal asks for one. This will usually be printed as a 'running head' at the top of the journal's right-hand pages.

The final abstract

Like the title, the abstract must convey as much information as possible when reprinted by an abstract journal or distributed by an information-retrieval service.

Abstracts may be classified as informative, indicative, or a mixture of informative and indicative (American National Standards Institute 1971). Write *informative* abstracts wherever possible: they are best for papers describing original research. They consist of succinct factual statements such as:

> A modified acetylene reduction technique was used for measuring microbial nitrogen fixation in four soil types in three areas of Europe. Maximum findings (μg/h in 1 g of soil) were: calcareous soil, Sweden 1.3–2.7; lime and loess, Germany, 0.07–1.6; mull, Andorra, 0. No fixation was registered in subsurface samples. Fixation in darkness, possibly bacterial, was 2–14% of values observed in the light.

Avoid the phrase 'X is discussed' in an informative abstract, for it is often a sign that the author is too lazy to sort out his conclusions (if any).

Indicative (descriptive) abstracts should be used only for reviews and similar papers. They contain general statements about the subjects dealt with in the article, i.e. its scope and coverage. Here is an example of an indicative abstract mistakenly written for a factual paper. Compare its information content with that of the abstract above:

> A method for determining the potential of microbial nitrogen fixation in soils under aerobic conditions is described. Four different soil types were studied: loess, lime, mull and calcareous. The soil samples were collected in three different areas of Europe.

Begin the abstract by stating the category (original article, case history, etc.) to which the paper belongs, if this is not obvious from the title. Describe the purpose of the investigation being reported but do not waste precious words by repeating or paraphrasing the title, since the abstract is always read in conjunction with the title. Indicate the methods used and summarize the results and conclusions. (Some journals prefer to have the most important results and conclusions first, then the supporting details, other findings, and finally methods.)

Keep to about 250 words for an article of 2000–5000 words and about 100 words for a short communication, according to the journal's requirements. Try to include all the basic information and to emphasize the different points in proportion to the emphasis they receive in the paper. Be brief, but state the purpose of the paper and the conclusions so that readers can immediately grasp the significance of your work. Write simply, as if you were describing the work for non-specialists. Do not add statements not made in the article itself. Do not keep rigidly to a recommended length if the subject matter demands a longer abstract for intelligible treatment.

The abstract should usually be compressed into a single paragraph, but split it into more if this is clearer for the reader. Some journals ask for numbered paragraphs or sentences in the abstract. Write complete, connected sentences and use the third person ('so-and-so was measured', not 'we measured so-and-so') unless this leads to clumsy or ambiguous constructions. This use of the third person, in contrast to the preferred use of the first person in the text, allows the abstract to be reproduced unchanged in secondary (abstract) journals. Avoid unfamiliar terms, acronyms and abbreviations—or, if you must use them, define them at first mention. Do not write footnotes to abstracts. Do not quote references to other work—or if you do, insert full bibliographical details in the abstract itself. Avoid tables, diagrams, equations and structural formulae if at all possible.

Selecting Index Terms

Many journals ask authors to supply about 10 index terms (key words or phrases; descriptors), suitable for use in the journal's subject index and in information-retrieval systems. In some journals, the index terms are *added* to the title in order to 'enrich' it and draw readers' attention to topics dealt with in the paper but not indicated in the title. In other journals, the index terms include words selected from the title. Find out which system is used by the target journal before you make your choice. Some journals specify that index terms are to be selected from a published thesaurus (for example, the Medical Subject Headings that are used in *Index Medicus*).

Assembling the Complete Manuscript

When you have written the final title and abstract and listed the index terms, check that all the parts of the paper are in the form specified in the journal's Instructions to Authors. Unless otherwise stated, items (1) to (9) described below should each start on a new page, as should the legends for the illustrations.

(1) **Title page**

On the title page put the title you have just written and the by-line—that is, the names of the authors (see p. 7) and the name and address of the institution (or institutions) where the work was done. Ask your co-authors exactly how they like their names to be written. Some people always write out one given name; others may leave out some initials. Decide early in your publishing career how you will write your own name for by-lines, and write it that way consistently to avoid confusion.

In writing the by-line, copy the style in the journal. Indicate any changes of address that have taken place since the work was done. Do not anglicize addresses in the by-line if your department or institution usually prints its address in a language other than English. If several authors from several institutions contributed to the paper, list them in such a way that it is clear who works where.

Add the name and the postal address of the author to whom correspondence (including proofs and requests for reprints) should go, unless the by-line shows this unambiguously.

(2) **Abstract**

Place the abstract on a separate page, after the title page, unless the journal gives other instructions. Comply with the journal's instructions on the lay-out of the abstract, and provide bibliographic and other information (title of paper, authors' names and addresses, name of journal) in the form stipulated. Leave room for the volume number, page numbers and date of publication to be inserted later.

Index terms

Put the index terms on the same page as either the title or the abstract, as specified by the journal.

(3) **Abbreviations**

List the abbreviations and their definitions on a separate page, after the abstract, if the journal asks for such a list. Keep this list short, and make sure that the abbreviations are acceptable (see p. 38 and Appendixes 2–4). List also any special symbols used, if the journal asks for these.

(4) **Text**

See that all the pages of text are present and correctly numbered in the top right-hand corner. Put the first author's name, or the short title of the paper if the journal requires this, at the top left of every page. Check the text again to ensure that spelling, symbols and abbreviations are consistent and correct throughout. If you numbered the paragraphs in the draft, remove the numbers now unless the journal uses them; check the headings and subheadings for usefulness and consistency. Make sure that you have referred to every figure and table at least once in the text, and indicated their approximate position by a circled note in the margin—'(Fig. 3 near here)'. Such notes are transferred to the margins of the galley proofs by the printer's reader; they enable the person who is making up the journal pages to lay them out most conveniently for the reader. Where you refer to other parts of the paper, write 'see Methods section' (for example) rather than 'see p. 10'.

In the text, use symbols for units of measure (e.g. A for ampere) whenever a number precedes them, but spell out the name of a unit if the number is spelt out, for example at the beginning of a sentence. The custom of writing out numbers at the beginning of sentences often leads to cumbersome phrases ('Eight microlitres per square centimetre') and some publishers have now abandoned it; if your journal retains this tradition, try recasting the sentence so that the phrase '8 $\mu l/cm^2$' falls somewhere in the middle. Avoid the double solidus because of its ambiguity: not W/m/K but W/(m K) or W m^{-1} K^{-1}. Use the decimal system for concentrations (0.5 mol/l); do not use fractions such as M/2. You may use '$^{0}/_{0}$' for percentage with an arabic numeral in the text, but avoid '$^{0}/_{00}$' for 'per thousand' because it is easily mistaken for $^{0}/_{0}$: give the unit of measure instead (5 $\mu l/l$) or write out 'per thousand' if there are no units. Never use '15 mg $^{0}/_{0}$': write '150 $\mu g/g$' or '150 mg/l', whichever is correct and in preference to '15 mg/100 g' or '15 mg/100 ml'). Mathematical expressions should whenever possible be written or transformed to allow typesetting on a single line.

(5) **Appendix**

If there is an appendix, place it either after the text and before the acknowledgements and references, or after the list of references, according to the practice of the journal.

(6) **Acknowledgements**

Place the acknowledgements on a separate page at the end of the paper, before the references, unless the journal prefers them as a footnote or elsewhere at the beginning of the paper. Keep them short. Ensure that names, initials, grant numbers and similar items are correct in every detail. When you thank individuals, make sure that they have agreed to the form in which you do this (see p. 25).

(7) **References**

Compile the references in the form required by the journal, as described below (pp. 54–57).

(8) Footnotes

Keep footnotes to a minimum, as mentioned earlier (p. 38), and place them according to the journal's instructions. Most journals ask for a separate page of footnotes, but some ask for footnotes to be typed on the appropriate page of text, either at the foot of the page or in the text immediately after the line in which the symbol calling attention to them appears, with lines ruled above and below to mark them off from the text.

(9) Tables

The tables will have been typed earlier (see p. 30). Now check once more that the observations noted in them are correct and consistent with the text, that totals and subtotals add up correctly, and that each table has its correct number and title. Identify each table in pencil with the name of the first author and (or) the title of the paper if the journal requires this.

(10) Illustrations

The illustrations that you drew or photographed, or sent to the technical artist or photographer (see p. 30), should be ready now. Check that they are all present and correct in every detail, complete with any necessary arrows, letters, scales or overlays. Write the first author's name and the number of the illustration lightly in pencil on the back of each illustration, as well as indicating which is the top. Check that the numbers and the information in the legends match the illustrations to which they refer.

Permission to reproduce material

If you have not yet received written permission to reproduce previously published tables, figures, text or unpublished work by other authors, obtain it now (see pp. 15–17).

Compiling the References in Final Form

Most journals use either a name-and-date system (sometimes known as the Harvard system) for citing references or a numbering system, in which the references are listed in the order they are first cited in the text. A third system is a hybrid numerical–alphabetical method, with the references arranged alphabetically by first author and then numbered. A fourth arrangement may be preferred in historical papers: references given in date order, the year being placed first. If you think this is appropriate, you should ask the editor whether he is prepared to accept it. Another method for references, rarely used in earth science or life science journals, is to place them as footnotes on the page on which the reference is given. This is expensive and unpopular with printers and it can also produce pages which consist mainly of footnotes, with just a few lines of text.

The first three systems all have their advocates, some of them fanatical, but you will have to use the system preferred by the journal for which your article is destined. The Harvard system seems to be gaining in popularity over the others, but each system has something to recommend it. For convenience and consistency most journals use only one of these systems.

References in the text

If the journal uses the name-and-date system, its instructions about references in the text should tell you how to cite (1) several papers published by the same author (with or without co-authors) in the same year (usually as 1974*a*, 1974*b* and so on, and *not* as 1974, 1974*a* . . .), and (2) papers with several authors (some journals ask you to write 'Smith *et al.*' or 'Smith and others' when the paper has more than two *or* more than three authors; other journals ask you to write out all the names the first time the reference appears). The instructions may also say that if you refer to a specific page of a book or article you should give that page number in the text only: 'This was reported by Smith (1974, p. 15).'

If the journal uses a numbering system for references, change the names and dates in your text to numbers at this stage (but first read the paragraph below about unpublished work). Write the numbers in the style used in the journal—usually either in parentheses or brackets on the same line as the rest of the text, thus (5), [5], or as small numbers above the line (superscript): [5], [5)], [(5)]. If you particularly wish to name an author, word the sentence accordingly: 'As Jones[5] has shown, this method . . .'. Write authors' names in capital letters only if the journal you write for insists on this.

With any system of citation, if you wish to give credit to others by referring to 'unpublished work' or to a 'personal communication', give the author's name and initials in the text—for example as '(X.Y. Smith, unpublished work)' or '(Y.Z. Brown, personal communication)' and do *not* make this into a reference for the reference list. Do not use 'in preparation', which implies a promise that may never be fulfilled, or 'private communication', which implies that you are violating a confidence. One reason for putting such references in the text instead of listing them with the references to published papers is that readers are immediately aware that the work has not been exposed to the critical appraisal of editors and referees, and that it is not readily available for critical study. Obtain written permission to cite unpublished work (see p. 16); the editor of the journal may ask to see this permission.

List of references

Call the list of references 'Literature Cited', 'References Cited' or 'References', according to the journal's style. Do not include any references that are not in the text. (If you are writing a review article or a textbook and wish to draw attention to other publications useful for background or further reading, make a separate list entitled 'Additional reading'.)

Only material already published or *accepted* for publication should be placed in a reference list. If you cite a paper that has been submitted for publication, do not include it in the reference list unless you are sure that you will be able to change 'submitted for publication' to the name of the journal plus 'in press' by the time your own paper is likely to be in proof. If you are not sure that a cited paper will be accepted, refer instead to 'unpublished work' in the text. Before the references are typed, remove from your collection of reference cards any that refer to unpublished work, or to personal communications: cite such work only in the text, as recommended in the previous section.

If you cite documents of limited circulation—for example, technical reports or typewritten or mimeographed papers issued by institutions, government agencies, or commercial companies on special request—you must be able to give the reader (in the list of references) full details of how to obtain them. If documents of this kind are not available on request, refer to them only if you can include in your text all the information necessary to support your argument, giving the reference as 'unpublished work', with the author's (or institution's) name. If such documents are classified as secret and their contents cannot be described, you will have to do without both the reference and the information in it.

For references that you have to cite at second hand, without having seen them for yourself, include a name and date or a number for the secondary source in brackets at the end of the primary reference—'. . . [cited by Braun 1957]' or '. . . [cited in ref. 3]'—and add the secondary reference in its appropriate place in the list. If you refer to an abstract, add this information in brackets too. If you refer to a paper in a language other than English, write '[in Russian]', or whatever language it is, at the end of the reference unless the bibliographic details make the language unmistakable. If you refer to reports of the kind that have copies deposited in many libraries, include reference numbers or other information needed for retrieving them. If you are in doubt about what bibliographic information to include, put in everything in full, as listed on p. 25: it is easier for an editor to delete details than to obtain them.

The different elements in references can be printed in an endless variety of ways, and before the reference list is typed you will need to arrange all the items on your reference cards in the order and with the punctuation favoured by the journal. You may underline—once—any items that will be printed in italics, but do not capitalize authors' names unless the journal asks for this to be done. Write the names of journals according to the system recommended by your journal.

There are two main systems for abbreviating the titles of journals for reference lists. One system—the 'International List' system, now the basis of a draft International Standard—is based on British Standard 4148:1970 and American National Standard Z 39.5–1969, which are in agreement. This is the system used by *Chemical Abstracts*, *Biological Abstracts* and *Index Medicus*. The correct abbreviations according to the International List system are most conveniently obtained from the *Bibliographic Guide for Editors and Authors* (BIOSIS *et al.* 1974), or from the most recent January issue of *Index Medicus*. Refer to the *International List of Periodical Title Word Abbreviations* (1970) for titles not named in these publications.

The other main system for abbreviating titles uses the abbreviations adopted by the *World List of Scientific Periodicals*, a three-volume publication last printed in full in 1963–1965 (covering journals published 1900–1960); this edition is supplemented regularly by the British Union Catalogue of Periodicals (BUCOP).

Some journals print the titles of all periodicals in full in reference lists. Check whether your target journal does this before you draw up your reference list. In any case, write out in full any journal title for which you cannot find an abbreviation. Do not abbreviate one-word titles or names of persons in a title.

Capitalize the elements of the title abbreviations according to the style of the journal. The International List system allows three styles:

Phys. Med. Biol.
PHYS. MED. BIOL.
Phys. med. biol.

The first style is used by most journals that abbreviate titles according to the International List system. Note that there is usually a full stop after *all* abbreviated words in a journal title, but no stop after complete words (such as *Acta*). You may omit accents and other similar marks in title abbreviations if the journal allows this, but do not change the spelling of any such words.

Examples of how to write out and type different kinds of references (using the name-and-date system of giving references and the International List system for abbreviating journal titles) follow.

(1) For journal articles:
Lourie, R. S., Layman, E. M., Jr and Millican, F. K. (1963) Why children eat things that are not food. Children 10, 143–146
The next example gives the title of the article in the original language and in English translation:
Akoum, G., Beucler, A., Gayet, N., Gall, G. & Brocard, H. (1973) Un poumon de champignonniste [Mushroom grower's lung]. Nouv. Presse Méd. 2, 1862–1863

(2) For a single-author book:
Cantarow, A. (1970) The Lifespan of Bees, 3rd edn., vol. 2, Nordic Publishing Company, Oslo
(If you need to give the total number of pages in a book, write this as 298 pp., not p. 298 or pp. 1–298.)

(3) For papers (titles omitted) in multi-author books with one or more editors:
Smith, P. D. (1969) in Biochemistry of Viruses (Brown, H. B., ed.), pp. 177–218, Associated Scientific Publishers, Amsterdam

(4) For whole documents such as books or technical reports (as for (3), but omitting page numbers):
Stubbs, A. (1965) Food Use and Nutritional Level of 1255 Texas Families (Bulletin No. H-B-1033), Agricultural Research Station, State College, Pennsylvania

(5) For theses and dissertations:
Smith, A. B. (1962) Population Growth in Fishponds, M.Sc. thesis, Dept. of Biological Economy, University of Leeds, England

Write out each reference in the correct form on a separate card. Then arrange the reference cards either in the order in which they are first referred to in the text (numerical system), or alphabetically (Harvard and numerical–alphabetical systems). Alphabetizing poses certain unexpected problems. Arrange double names and names containing connectives such as 'van', 'de', etc. according to the policy of the journal. If no policy is apparent, write the names according to the custom of the country in which those authors live, or according to a self-consistent system in which all names beginning with 'van', 'Van', 'von', etc. are listed under 'v' whether the first letter is capitalized or not.

The most important thing is to be consistent and write the names the same way in the text and in the reference list, so that the reader can find the references where he expects to find them.

Name an organization (for example, Ministry of Education) as the author of a report or similar publication when no individuals are named, and add the country's name to the bibliographical details if this is not obvious from the place of publication. List an unsigned article under 'anonymous' (abbreviate this to 'Anon.' in the text).

Even when the same authors' names or journal titles occur in successive references, write them out in full; do not use *ibid.* or similar shorthand tags like *idem*, *loc. cit.*, and so on. Too often these terms are ambiguous, and if a reference is later deleted they may become inaccurate.

Checking the reference cards

Now check the details on the reference cards or rough list against the original documents, or against reprints or photocopies of them. Except in the rare cases when a journal is paged separately for each issue rather than consecutively through the whole volume, delete the issue numbers on the reference cards before the list is typed.

Finally, before the paper is typed, check the reference cards or rough list against the text. With the Harvard system, check that the names in the text are spelt in the same way as on the reference cards, that they are the same names in the same order, and that the dates are the same as those on the cards or rough list. With the numbering system, make sure that the numbers in the text refer to the publications to which you intended them to refer—listed or arranged on your reference cards with numbers matching those in the text. For a 'Reference list' (but not for a list of 'Additional reading') make sure that every reference is referred to in the text, and that every reference in the text has a corresponding reference card or appears in the rough list.

Although a skilled typist will be able to look after many of these details if you first explain the system used by the target journal, typing the references will be quicker and easier if you follow the above recommendations, writing every detail of punctuation, capitalization and italicization in exactly the form the journal wants, and arranging the cards or list yourself.

Typing

Type the whole paper, or have it typed, according to the recommendations in chapter 6, but tell your typist to observe the journal's instructions in preference to ours if the two sets are in conflict.

Correcting Typescript

Read and check the typescript word by word when it is ready: if possible, get someone to read the original aloud while you make any necessary corrections on the top copy and on all other copies. Do *not* use proofreaders' marks for correcting typescript. One reason for having everything typed double-spaced is precisely to allow words or characters to be inserted or substituted between the lines, where they are most easily seen by the compositor who sets the type at the printer's. The compositor can deal perfectly well with clear deletions, or insertions between the lines, but not with long insertions typed at the bottom of the page or with words or phrases marked to be moved to another page. If one or a few lines have to be corrected, type them on a slip of paper which can be pasted over (not pinned to) the original with a non-aqueous glue. Extensively corrected pages must be retyped. Of course, if the journal is printed by photo-offset direct from authors' typescripts (see p. 59), whole pages in the final version will have to be retyped even when the corrections are minor ones.

Obtaining Final Criticism

Obtain a final critical review either from the head of your department or from an experienced colleague (possibly working elsewhere) who is doing similar work but who has not been involved in preparing the manuscript. Ask him or her to point out errors of detail as well as making larger-scale comments. While you wait for criticism, put away the other copies of the typescript. Try to achieve a sense of detachment from the work you have described, so that when the critical comments arrive you can re-read the paper objectively with a fresh eye. Make any final changes neatly on the typescript, or have the relevant sections or the whole paper retyped, before you send anything to the journal. The mechanics of submitting the paper are discussed in chapter 7.

Chapter 6

Typing

Editors and referees get their first impression of a paper from its physical appearance. The way it is typed is therefore important, even though the intrinsic merits of the work described should in the end be the primary basis for its acceptance or rejection. A neat, clean and accurately typed article, on good paper, with a carefully laid-out title page attracting the eye, will make the editor's (or referee's) work much easier; he or she will be at least slightly biased in the author's favour at the outset. Later, these same qualities will greatly help the copy-editor and compositor, and there will be fewer mistakes in the printed version. If the journal is photographically reproduced directly from authors' typescripts a high standard of presentation is obviously essential; the journal may also ask for typing to be done on special paper. Corrections at proof stage are very expensive with this method.

The rest of this chapter is therefore addressed to the person—perhaps yourself—who is going to type the later drafts or the final version of your article. Even if you are *not* doing your own typing, you should read these recommendations. If our recommendations differ from the requirements of the journal, point out the differences and give the typist a recent copy of the journal and its instructions, as well as calling attention to this chapter. Alternatively, make your own list of instructions about how to type a paper for the journal. Find out how many copies of the paper should be made: *as a minimum*, you will probably need two for the journal, one for each author, and an extra one for your files. If you want the paper typed by a certain date, say so, but do not hurry the typist unnecessarily: a good-looking and accurate typescript takes time to type. Remember to tell the typist when you have reached the final version—the one that will be submitted to the journal.

Format and Materials

Use white paper, A4 size (210 × 297 mm) for the top copy, typing on one side of the paper only (many American journals specify paper measuring 8½ × 11 inches, or about 215 × 280 mm). The paper for the top copy must be strong and of good quality, because the typescript will be handled often by many different people. Make two or more extra copies, as specified by the author; these may be either clean carbon copies on a lighter-weight paper, or clear reprographic copies. Change a spool of typewriter ribbon as soon as the type begins to look faint. Type *everything*, including tables, legends, references and footnotes, in *double* spacing (one typed and one blank line about every 8.5 mm—that is, three typed lines to the inch), but leave more space than this above and below headings, equations and formulae. Leave wide margins (25 mm or more) all round, with at least 30 mm on the left so that the editor has space for instructions to the printer and—sometimes—queries to the author.

Pagination

Arrange the pages of the manuscript from which you are working in the following order (items marked (1) to (10) will be started on a new page in your final typescript):

(1) title page, with title, authors' names and addresses, index terms if required here, short title, and mailing name and address (see pp. 49–50);

(2) abstract and index terms, if the latter are not to go on the title page (see p. 50);

(3) abbreviations or glossary, and list of special symbols used, if any of these have to be listed separately;

(4) text (see pp. 50–51);

(5) appendix, if any (or place before the acknowledgements and references, depending on the requirements of the journal);

(6) acknowledgements (see p. 51);

(7) references;

(8) list of footnotes, if these have to be placed on a separate page (see p. 52);

(9) tables, with each table starting on a new page; and

(10) legends for the illustrations.

Type the first author's name, or the short title if specified by the journal, at the top left of every page (except the title page) so that the sheet can be identified if it gets separated from the rest of the typescript. Number the pages of the typescript consecutively at the top right (this being the easiest place for checking the numbers). Begin the consecutive pagination with the title page and continue through appendixes, references and footnotes. Continue the numbering in pencil on the legends unless instructed otherwise.

Insertion of extra pages

If extra pages are added when you have finished typing a long paper, number new pages inserted after page 5, '5B', '5C' etc. and renumber page 5 as '5A'; at the bottom of the last inserted page write 'p. 6 follows'. If a page is removed, add its number to that on the preceding page: thus, if page 6 is removed, make page 5 into '5 and 6'. Never place insertions on the back of a page or on slips of paper stapled or pinned to a page; if a correction or insertion is too long to be written neatly above the relevant line, or typed on a slip of paper which can be pasted over the original, retype the whole page and add the appropriate letter to the page numbers if the retyping produces an extra page.

Typing the Text

You will probably have to retype the text of most papers several times. Do not be irritated or upset by this: few authors achieve perfection in the first draft and they cannot produce a satisfactory final version without a fresh typescript to work on from time to time.

Since it is particularly important that the version to be submitted to the journal is in perfect form, ask the author whether the material given to you for typing is still a draft, or the final version. At the final-version stage, try not to let a time limit fluster you into doing poor work.

Try not to break words, *especially hyphenated words*, at the end of a line: printers are trained to leave out hyphens at the ends of lines in manuscripts and make the letters on both lines form a single word. On the title page, as well as in the text, capitalize only those words or letters that are normally written in capitals. Underline (once) words that are normally printed in italics, such as names of species; do not underline the title of the paper, except for words or characters that are to be italicized, and do not use double underlining. Too much underlining makes it difficult for the editors to instruct the printer about italics and capitals.

Distinguish by means of a pencilled note the figure 1 (one) from the lower-case letter l and from the capital letter I where this is not clear from the context; similarly, distinguish x (letter) from × (multiplication sign) and O (capital letter) from 0 (zero). Do not type a 'u' for the Greek letter μ, or 'X²' for χ^2; do not type the name of a Greek letter such as alpha, beta, gamma, lambda, etc., when correct usage requires the letters α, β, γ, λ, etc. Put these in by hand if they are not on your typewriter, making sure that they come out clearly on every copy of the page. (If you need these characters often in your work, have them put on the typewriter instead of fractions or other rarely used symbols.) Greek letters should be identified in pencil in the margin.

Headings

The author's text should preferably have not more than four orders of headings (see p. 21)—1st, 2nd, 3rd and perhaps 4th order, marked as such in pencil in the margin. Type the headings in the form agreed with the author. Capitalize only those words or letters that are normally written in capitals, except in any headings that have to be typed all in capitals. Do not letterspace headings or place full stops after them. Do not underline headings except when the journal instructs you to do so, or when it is essential for clarity in reading; never underline headings that are all in capitals.

References in the text

Type references in the text according to the author's instructions and those of the journal—that is, type *either* numbers *or* names and dates (see p. 52). If the journal uses a numbering system, read the Instructions to Authors or look at an article in the journal itself to see how the number is printed: it will usually be on the line in parentheses—(5)—or brackets—[5], or it may be superscript (raised half a line) in the form ^{5}, $^{5)}$ or $^{(5)}$. If the author has been inconsistent in the way he has written references in the text, ask him or her about this before you type any more of the paper.

Numbers

Type figures, not words, for all measured quantities ('9 mg'). In descriptive text spell out whole numbers from one to nine ('eight dogs'); use figures for 10 and over, except at the beginning of a sentence, but type '4 out of 15 dogs' (that is, do not mix figures and words referring to the same noun in a phrase or sentence). In English you must use a decimal point on the line, *not* a comma, in decimal quantities (21.9). Place a zero before the decimal point for all quantities of less than one (0.695; –0.28). When numbers consist of more than four digits, leave one space between each group of three, counting in either direction from the decimal point:

5 213 504 23.587 62

For numbers with only four digits to the left or right of the decimal point do not leave a space (5123; 0.5876) except when you are aligning such numbers with others of more than four digits, in the column of a table:

5123	but	5 123
3125		60 486

Abbreviations

The author should have ensured that abbreviations in the text (*a*) are kept to a minimum and (*b*) are either internationally accepted ones or are defined at first mention. If the abbreviations seem inconsistent, or if conditions (*a*) and (*b*) are not met, point this out to the author. Once an abbreviation has been introduced (and properly defined) in the text, the spelt-out form should not appear again; if it does, there may be a good reason, or it may be an oversight: ask the author.

Quotation marks

If the author has quoted passages or phrases from other people's work, copy these exactly—even mistakes in the original, if any. If you have both single and double quotation marks on your typewriter, type single quotation marks, not double, unless you have been asked to do otherwise. Use double quotation marks, if they are on your typewriter, for a quotation within a quotation. Or you may indent long passages and omit quotation marks altogether. Quotation marks may be placed round a word used in an unusual way, or round newly coined words; put them in the first time the word appears in a paper but not the second or any other time.

Hyphens, en dashes, em dashes and minus signs

A typed hyphen may be printed as either a hyphen or an en dash (that is, a dash the width of the letter 'n' in the type face being used for the journal). An en dash is often printed to indicate a single bond in chemistry (C–H), a range (20–25 mg), distance or movement (Paris–Brussels), or a combination (gas–liquid). There is no need to type these two signs differently, but you should distinguish a minus sign from either a hyphen or an en dash by leaving a space on each side of a minus sign (10 – 3 = 7) unless it indicates a negative number (3 – 10 = –7). Similarly, leave a space on each side of an 'equals' sign (c = 6) to distinguish it from a double bond (C=O) in chemistry. A dash in the middle of a sentence—used when the author wants to separate a portion of the sentence dramatically, as here—may be indicated by a double hyphen without spaces. It is printed as an em dash (that is, a dash the width of the letter 'm').

Mathematical formulae and equations

Type mathematical work exactly as written by the author, as far as the symbols on the typewriter allow (and see pp. 51, 61 and 62). Some mathematical expressions may have to be written in by hand; if so, the author should tell you where, and how much space to leave.

Typing the Reference List

The author will have given you a series of reference cards, or perhaps a handwritten list of references, in the order in which they are to be typed. If the details have not been written in exactly the way required by the journal, change them carefully as you type. Read the relevant portion of the journal's Instructions to Authors, or look at a list of references in a *recent* issue of the journal, or do both, to see how the references must be done (and see pp. 54–57).

Remember to double-space all references. For those that are more than one line long, indent the second and any succeeding lines by two or more spaces. Type the authors' names in full each time, even when the same names appear in successive references; do not use '*ibid.*' or similar terms, or put dashes or ditto marks (unless this is the journal's style). Underline the parts that you *know* will be in italics when printed.

If the journal uses a numbering system for references, check whether the number at the beginning of each reference should be typed above or on the line, whether it is followed by a full stop, and whether it is in parentheses.

When you have finished typing the reference list, check each item in it carefully. Every letter, punctuation mark, space and digit must be correct.

Typing Tables

Number tables with arabic (1, 2, 3 . . .) numerals unless the journal prefers roman (I, II, III . . .). Each table number should have a title after it; do not underline the title except for words that are normally italicized. If the title occupies more than one line, double-space the lines. Type the title above the table, not on a separate page from the table.

There may be a line or two of information under the title or below the rest of the table. Remember to double-space this material, like everything else in or around the table; do *not* squeeze it into the smallest possible space. If necessary, continue typing the table on a new page. If tables are typed in a cramped way they cannot be clearly marked for the printer.

Do not type more straight lines above, below, or within tables than are required by the journal's style. Look at a table in a *recent* copy of the journal for guidance on this. In particular, do not insert vertical lines between columns unless you are asked to do so.

Column headings in tables are one place where you may, if necessary, split words and type single-spaced, although it is still better to avoid doing either if you can. In general, type units of measure once, in the column heading, rather than repeating them all down the column: but ask the author about this.

If the table is too wide to be typed on one sheet of A4 paper (turned sideways if necessary), it is probably too wide for the journal to print. Before you type a gigantic table on many sheets of paper which would then have to be glued or taped together, ask the author whether the table can perhaps be redrafted.

Align numbers in the body of the table on the decimal point, expressed or implied. If $\pm$ or $=$ signs are used, align first on the $\pm$ or $=$ and secondarily on the decimal point:

$$\begin{array}{r@{}l@{\;\pm\;}r@{}l} 68&.1 & 1&.5 \\ 234& & 61& \\ 0&.29 & 0&.03 \end{array}$$

When a table is too long for one page, type 'Contd' on the bottom of the first page and 'Table 1, page 2' at the top of the next, then repeat all the column headings exactly as on page 1 before continuing. Do this again for each continuation page.

Type the explanatory notes for a table immediately under that table, *not* on a separate page (but continuing on a new page if necessary, with the same notations of 'Contd' and 'Table 1, page 2' as above). Each explanatory note will be identified by a superscript letter or number (^{a}, ^{b}, ^{1}, ^{2} etc.) or by a symbol. When the table has been typed, make sure that all the signs have been inserted, by hand when necessary, and that each sign in the explanatory notes appears in the table itself (caution: it may appear more than once in the table). Type notes to tables double-spaced, like everything else.

Do not confuse explanatory notes to a table with footnotes to the text. The latter *are* often typed as a list on a separate page.

Typing Legends for Illustrations

Type the legends (captions or titles, plus other descriptive material) for illustrations as a double-spaced list, headed 'Legends', on a page or pages at the end of the manuscript. See how the legends in a copy of the journal have been printed, and if possible have a copy of your author's illustrations before you. Each legend will begin 'Fig. 1 (2, 3, etc.)' or 'Figure 1', according to the journal's style. After 'Fig. 1' comes a title, occasionally underlined, and then the text of the legend. This may include letters or symbols that have to be put in by hand: check that they correspond to those in the illustration itself. Make sure that spelling, abbreviations and so on are consistent between each legend and the corresponding illustration, and between different legends.

Checking

Read the whole paper through before the author corrects it and has it critically reviewed (p. 58). Make sure that nothing has been left out. Check especially that all numbers in the text, tables and legends are correct. Re-read this chapter quickly, using it as a checklist of what must be present (p. 60) and how it should look, for *every* copy of the typescript.

Chapter 7

Submitting the final version

When you (the author) have marked the final corrections on all copies of the typescript, write a short covering letter to the editor. If the work you are reporting is part of a series, or if it is closely linked with an earlier paper in the same journal or elsewhere, refer to these earlier publications in your letter. Some editors also like to receive relevant reprints. If the work, or a part of it, has been reported before (for example, in a report in a scientific newspaper or popular journal) in any form you must tell the editor about the earlier version and send him a copy of it.

If your tables or graphs are extremely large or numerous tell the editor whether you are willing to have this material, or some of it, sent to an information depository. For a journal that publishes several categories of article, name the category to which you think your paper belongs. If—and only if—your field is a highly specialized one, the editor may want you to suggest the names of possible referees. You will probably find guidance on these points in the journal's Instructions to Authors.

Give the address and telephone number of the author to whom correspondence and proofs should be sent, when it is not yourself.

If the paper describes research investigations in man or on animals with a central nervous system, enclose a copy of the authorization or approval of this work issued by the ethics committee of your institution. Where relevant (see p. 8), assure the editor that experimental animals were well treated and cared for.

Enclose copies of any letters ('releases') from authors or copyright-holders giving you permission to cite unpublished work or to reproduce tables, illustrations, or more than 50 words of previously published material (see pp. 15–17). You need not send copies of permission to acknowledge other kinds of help.

Finally, check that all pages of each copy of the manuscript are present, in the order listed on p. 60, and that each copy includes a complete set of tables and illustrations, or copies of them (see also the checklist on this page and the next). Place thin cardboard or other protective covering round the illustrations if necessary. Send the top or 'ribbon' copy, plus any extra copies stipulated, to the journal. Even if you send the paper by registered post or by a similar certified method, as required by some journals, keep at least one complete copy for yourself, and give one to each of your co-authors.

Although most journal offices use a printed card to acknowledge receipt of the manuscript, it can do no harm to enclose a self-addressed postcard which the editor can use for such acknowledgement.

It will probably be several weeks before the editor makes a decision about the paper. Do *not* submit another copy of the paper to a second journal even if you become impatient at what seems a long time before you receive a decision from the first journal. You could be in considerable difficulty if both of them accepted the paper (see p. 4). Allow six weeks or more to pass before you write to ask whether a decision has been made about accepting or rejecting your paper. The editor's reply will determine your next step, as described in chapter 8.

List of Items to be Checked Before Submission of the Paper*

Make sure that you have:

(1) Numbered the text pages consecutively, beginning with the first or title page.

(2) Numbered your tables (typed separately from the text, and not more than one on a page) consecutively in the order in which you want them to appear in the journal.

(3) Read the title and headings of each table objectively to determine whether the table can be understood without reference to the text.

(4) Searched the text for references to tables to make certain that each table is referred to and that each of the references is to the appropriate table.

(5) Indicated by a marginal note a place for each table.

(6) Examined your text, tables and legends to make certain that each reference cited is accurately represented in the reference list.

(7) Examined your reference list to make certain that each work listed there is accurately referred to in the text, tables or legends.

* Adapted, with permission, from *CBE Style Manual* (Council of Biology Editors 1972), pp. 170–171.

(8) Examined each item in the bibliographic section for accuracy of dates, wording, spelling and other details (by comparing each item with the original article or book, or with a previous draft that has been carefully checked with the original sources — *not* with your rough notes or with citations by other authors).
(9) Prepared adequate legends for all illustrations (legends double-spaced on a separate page or pages, *not* typed on or below the illustrations—unless you have supplied a separate set of legends as well).
(10) Made certain that illustrations are numbered consecutively in the order in which you want them to appear in your article, that each of them is referred to at least once in the text, and that each reference in the text is to the appropriate illustration.
(11) Indicated by a marginal note a place for each illustration.
(12) Examined the text for references to footnotes, making sure that each such reference has a footnote and that each footnote is referred to in the text.
(13) Reconsidered the appropriateness of your title and abstract and your index terms (if any).
(14) Reviewed the special requirements of the journal to which you are submitting your manuscript and made certain that you have met them.
(15) Carefully read your final typescript at least twice, once checking against the pages from which it was typed, and the second time—preferably on a different day—just reading it straight through.
(16) Prepared as many copies of your text, tables and illustrations as are required by the editor to whom you are submitting your manuscript, by your co-authors, and by the institution for which you work.
(17) Kept for your files a complete copy of your manuscript and accompanying material (corrections and insertions included).
(18) Enclosed copies of releases for material requiring releases.
(19) Included on the first page of the typescript the address to which letters, proofs and requests for reprints should be sent.

Chapter 8

Responding to the editor

The editor's reply will state either that the article has been accepted or rejected, or that it will be accepted or reconsidered if you revise it in ways suggested by the editor or referees.

Acceptance Without Revision

It is unusual for a paper to be accepted without queries of any sort. If you are lucky enough to have this happen, you need take no further action until the proofs reach you, except perhaps write a note thanking the editor.

Rejection

If you receive a rejection letter, study the reasons given and follow one of these four courses of action:

(1) If the editor says the article is too specialized, not specialized enough, or outside the scope of the journal for some other reason, your best course is to send it to another journal, first modifying the style to comply with the instructions of that journal. It is useless to argue—especially as some editors use 'The paper does not lie within the scope of the journal' as a euphemism for 'We don't like this article'. Another euphemism for outright rejection is 'The journal is very pressed for space'.

(2) If the article is rejected because it is said to be too long and in need of changes, consider shortening and modifying it according to the criticisms put forward—and then submit it to a different journal, again with the style appropriately changed. If the editor of the first journal had wanted to see a shorter version he would have offered to reconsider it after revision. If you are uncertain whether this is what the letter means, ask the editor.

(3) If the editor thinks the findings reported are unsound or that the evidence is incomplete, put the paper aside until you have obtained more and better information, unless you are sure that the editor and his advisers are wrong.
(4) Consider contesting the decision only if you honestly think, after considerable reflection and at least one night's sleep, that the editor and referees have made a superficial or wrong judgement. In this case write a polite letter explaining as briefly as possible why you think the editor should reconsider his decision; enclose a copy of the manuscript if the editor no longer has one. Do not telephone, and never write contesting the decision unless you have thought it over extremely carefully.

Revision Requested

The letter you are most likely to receive is one asking for specific changes to be made before the article can be either accepted or reconsidered. If the editor says that the paper will be *accepted* if the changes are made, consider the suggestions carefully, and if you agree that they will impove the paper, modify or rewrite sentences or sections as necessary. Retype any heavily corrected pages before you return the paper to the editor, but enclose the original corrected pages as well as the retyped copies. In your covering letter sent with the revised version, thank the editor and referees for their help and enclose a list of the substantial changes made in response to their suggestions; if you have rejected one or more of the recommendations, explain why.

If the editor has merely offered 'further consideration' and the changes suggested are major ones, you will have to think hard about whether the effort is worth while. You may eventually decide that the paper is better as it is, and proceed to try another editor in the hope that he will agree with you.

General

If your article is rejected, or accepted only on condition that you revise it drastically, *do not* succumb to the temptation to enter into acrimonious correspondence with the editor. Criticism will nearly always have been made dispassionately, for the sake of the journal's reputation and the furthering of science, not with the aim of discomfiting you or denigrating your work. Even if the criticisms are mistaken, anger will not help matters. Most referees are impartial in their assessment of papers submitted to them, and too busy to spend time and effort inventing trivial objections to your paper. Consider their recommendations seriously, therefore, and try to make constructive use of the criticism so disinterestedly provided for you.

Above all, do not try to guess the identity of anonymous referees. You will almost always be wrong, and may harbour resentment against a completely innocent colleague for the rest of your career. Occasionally, an editor will put an author in direct contact with a referee (after obtaining the latter's permission) in the hope that they can between them resolve a complex or highly specialized problem.

Chapter 9

Correcting the proofs

First Proofs

The proofs that you receive for correction will usually be either galley proofs (long sheets not yet divided into pages) or paged galleys: correct these as described below. However, if the journal is printed by photo-offset from typed pages you may receive a new typescript with a transparent overlay: correct this lightly in pencil on the overlay, in the way you would correct typescript rather than printed proofs, making sure that the overlay is lying flat on the page as you do this. If the journal is phototypeset by computer, you may not receive any proofs at all; you will have to be sure that the typescript you submit is perfect, and trust that the proofreaders at the printer's or in the editorial office will make certain that the printed version corresponds exactly to your original copy. You may sometimes receive an edited copy of your typescript. If you do, check it carefully, as if it were a set of proofs, and correct it in ink that is noticeably different from that used by the editor; the editor may check the proofs against this corrected typescript instead of waiting for you to send proofs back later.

The editor usually wants the proofs back almost before they reach your desk, so deal with them immediately you get them. You will need to read the proofs at least twice. If you think you will be away when they are due, ask a co-author or colleague to check the proofs for you, or consult the editor about what should be done.

First ask someone else to read aloud from the original typescript if this has been returned to you, or from your file copy (which of course will not show the editor's changes), while you verify all numbers in the text, tables, illustrations, legends and references, and check the proofs for accuracy of spelling, especially of proper names and unusual words.

See that no sentences or paragraphs in the text have been left out or repeated, and that the headings and subheadings are correct. Check that the tables are correctly numbered and positioned (or planned for positioning). Make sure that all the elements of every reference in the list are correct.

Ensure that illustrations have been accurately made, printed the right way up, and correctly numbered to correspond to the legend of the same number, and that a reference to each illustration appears at least once in the text. On illustrations proofed separately from the text, write the figure number and indicate which is the top; in the margin of the text, encircle a note—'(Fig. 2 near here)'—indicating where each illustration should be printed. Check any magnifications given in the legends. If the journal is printed by an offset process, the 'bromides' or blueprints you may get will be photographic reproductions of the illustrations, not proofs made from the plates that will be used for the final printing; the printed versions you eventually see in the journal will be quite different in quality, although what they represent ought to be the same as on the bromides. If letterpress printing is used, do not expect half-tones to have the quality of the final product unless proofs are supplied on the same quality paper as will be used in the journal itself. Do not add new material to an illustration or make any changes in it unless you can see that it contains a serious mistake. Alterations at this stage will probably mean that a new block has to be made. If corrections are essential, return the illustration to the editor with the corrected proofs. A few journals re-draw or re-letter all, or nearly all, the illustrations submitted by authors; check the proofs particularly carefully where you can see this has been done, as errors are sometimes made by the draftsman.

The second time you go through the proofs, read them for comprehensibility and for accuracy of the statements made. Make any essential changes, but remember that the place for improving your style or making minor alterations was the typescript, not the proof. Do not delete material unnecessarily, and do not add new material without telling the editor; if you have acquired relevant new information that you think is important enough to be included in the article, write an addendum to go at the end, between the text and the references. This is usually headed *Note added in proof.*

Make corrections so that they produce the fewest possible changes in the lines of text. This will be cheaper for the journal—and for you too, if the journal passes correction charges on to authors. If you add new material, apart from an addendum as just described, try to add it at the end of a paragraph or as a new paragraph; alternatively, delete as many characters as you add (count each space as one character). Similarly, if you delete material, try to add the same number of characters and spaces. Of course, none of these manoeuvres should be at the expense of intelligibility or scientific validity. Resign yourself to the fact that the editor may ignore changes that go against the 'house style' of the journal, or any that you have made for stylistic reasons but without any significant gain in intelligibility.

Marking the corrections

Mark all corrections and instructions in the *margins* of the proofs, with corresponding marks in the text itself (see below). Neither the editor nor the printer's compositor will look within the text for a correction unless they see a mark in the margin drawing their attention to a requested change. Write neatly and legibly, in ink of a different colour from any already used by the printer or editor on your proofs. Similarly, if you are asked to distinguish printer's errors from author's changes, do so by using differently coloured ink or in any other way specified in the instructions.

You do not need to know or use the whole battery of symbols used for proof correction: your corrections will normally be transferred to a master set of proofs in the editorial office and all you need to do is make your intentions clear and easy for the editor to see. However, a short set of correction marks is given at the end of this chapter (pp. 78–79). If the correction you want to make is not listed there, write your instruction in plain language in the margin and circle it, write the correction or addition beside the instruction, and put an appropriate mark in the text to show where the new or corrected material is to go. Always circle instructions that are not to be printed.

The proofreader's mark *in the margin* tells the printer what the change is—whether it is a character or word to be inserted, a change of type face, or the deletion, closing up or moving of type as set. Place these marks close to the part to be changed—in the left or the right margin—and at the same level as the line to which the change applies. If there are several changes in one line, arrange them in order from left to right; if you are dealing with a British printer, separate each marginal change from the next by a diagonal line or stroke (/). The system of distinguishing each mark in the text by a flag or ring on each stroke works well with most European printers but it is not used in Britain or America.

The proofreader's mark *in the text* shows where the change written in the margin is to be made. The text mark will usually be an insert mark or caret (⅄), a line striking out a character or word(s), or some kind of underlining (to indicate italics, small capitals, capital letters, or restoration of material that you have deleted by mistake). The example overleaf shows how a few of the more common corrections can be indicated in the text and in the margins.

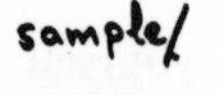

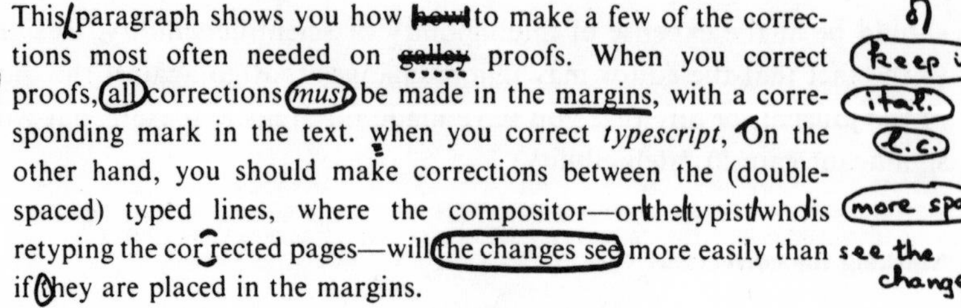

This paragraph shows you how how to make a few of the corrections most often needed on galley proofs. When you correct proofs, all corrections must be made in the margins, with a corresponding mark in the text. when you correct *typescript*, On the other hand, you should make corrections between the (double-spaced) typed lines, where the compositor—orthetypistwhois retyping the cor rected pages—will the changes see more easily than if they are placed in the margins.

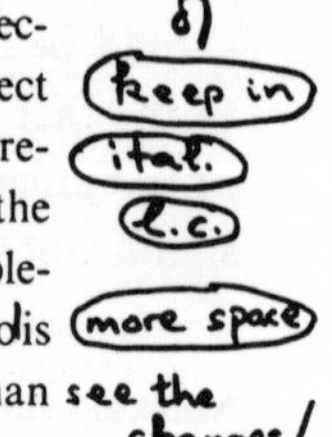

When corrected, the above paragraph reads:

> This sample paragraph shows you how to make a few of the corrections most often needed on galley proofs. When you correct proofs, **all** corrections must be made in the *margins*, with a corresponding mark in the text. When you correct *typescript*, on the other hand, you should make corrections between the (double-spaced) typed lines, where the compositor—or the typist who is retyping the corrected pages—will see the changes more easily than if they are placed in the margins.

If a correction affects two lines or more, type it and tape the slip of paper to the left margin near the appropriate place; do not use a pin or paper clip. Alternatively, type the passage directly onto the top or bottom of the galley proof page. In either case, make it clear where the insertion is to go.

Mark each change as you come to it: do not write, for example, 'Print the word Sun with an initial capital throughout'. The printer will make only the corrections that he sees marked in the margins of the proofs. If you discuss or list changes in a letter to the editor, you must mark them on the proofs as well as in the letter.

When you receive your proofs, there may be question marks in the margin indicating doubt about a particular passage, and more detailed queries from the editor or printer's proofreader. If you want the material to remain as already printed, put a line through the question mark but do not erase or obliterate it. Answer the queries unambiguously.

If you received two sets of galley proofs, mark the second set in the same way as the first and keep one set for your file. Initial and date both sets. Return one corrected set as quickly as possible—by airmail when necessary—to the address given in the letter or form sent to you with the proofs. If you are ordering reprints, send your order as instructed by the journal. But do not let any consultations with co-authors or your institution over the size of the reprint order delay your return of the proofs: send these on at once and explain that the reprint order will follow shortly.

Second Proofs

Authors rarely receive second (page) proofs. If you do receive them, check first of all against your copy of the corrected galleys to see whether the changes have been correctly made. Check the rest of the line in which each change was made, and the lines above and below it, since moving a small piece of type sometimes disturbs the type near it. Check whether the tables and illustrations have been placed as conveniently for the reader as the circumstances allow. (The necessity of having a journal printed economically, on pages of a fixed size, means that the position of these items will not always correspond to what you had visualized as ideal). Check the running heads or footlines. Examine the first and last words on each page for missing letters. Be particularly careful about examining the title, by-line, and everything else that precedes the text. If there is time, read the whole article through once more. *Absolutely* no changes other than the correction of major errors are allowable at this stage. Mark corrections in the same way as for first proofs. Prompt return of the page proofs is usually even more desirable than with the first proofs.

You may wonder why it takes so long for your paper to get into print. This diagram, illustrating what happens to it between submission to the editor and publication, may suggest some reason for the delay.

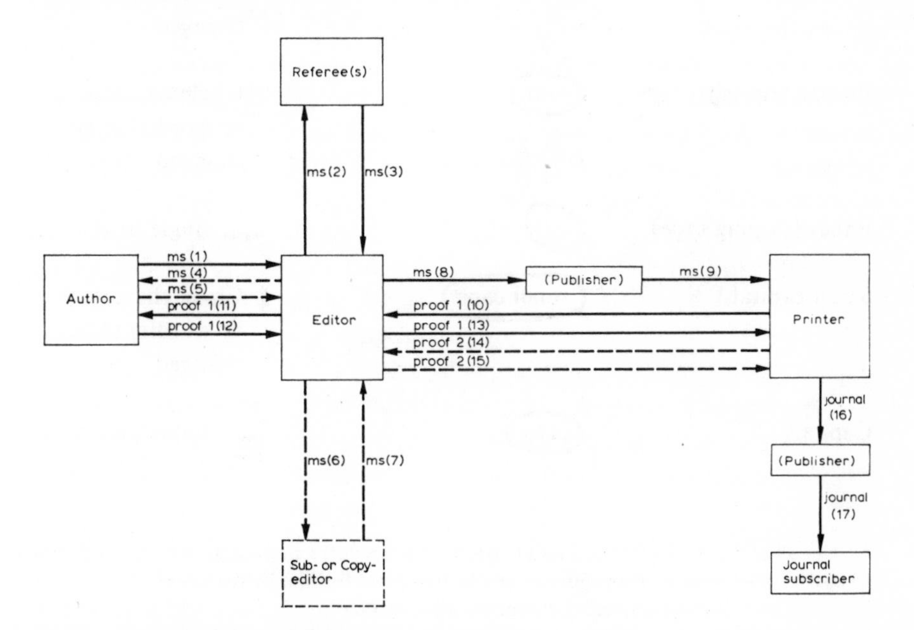

Fig. 9. Stages in getting a submitted paper published.

Proof Correction Marks or Instructions to the Printer*

Meaning	*Mark or circled instruction in margin**	*Mark in text†*
Insert	⅄ at end of new material	⅄
Delete	ꝺ	/ or — through character(s) or word(s) to be deleted
Substitute new material	Write the new character(s) or words [not circled]	/ or — through character(s) or word(s) to be deleted
Keep in material wrongly struck through	(keep in)	---- or .... below material to be kept in
Bold (heavy) type	(bold)	‡ Circle character(s) or word(s) to be changed
Roman (upright) type	(rom.)	‡ Circle character(s) or word(s) to be changed
Italics (sloping type)	(ital)	— single underlining
Small capitals	(small caps.)	‡ Circle character(s) or word(s) to be changed
Capitals	(caps.)	≡ treble underlining

* Not all of the circled instructions are standard proof correction marks, but they will either be easily understood by the printer or can be translated for him by the editor.

† These are all standard marks except the four marked ‡.

‡ Non-standard marks.

Meaning	*Mark or circled instruction in margin**	*Mark in text*†
Lower-case (small) letter(s) instead of capital(s) or small capital(s)	(l.c)	Circle the letter(s) to be changed
Increase space	(more space)	ǀ between characters requiring space
Decrease space	(less space)	⊂
Close up (no space)	(close up)	⁀/‿
Transposed characters or words	Write the correct sequence [not circled]	‡ Circle the characters or words to be changed
Damaged, dirty or inverted characters, or characters of wrong type face	X	Circle the characters
New paragraph	(new para.)	⌐
No new paragraph	(run on)	⊂~
Superscript character (above line)	Write the character [not circled] and add (superscript)	/ or ⅄
Subscript character	Write the character [not circled] and add (subscript)	/ or ⅄
Indent	(indent)	⊣
Align	(align)	‖ or =

Appendix 1

Steps in writing a paper

Step		*Pages*
1	Assess your work: decide what, when, and where to publish. Refrain from duplicate publication, and define your purpose in publishing	3–6
2	Obtain and read the Instructions to Authors of the journal chosen	6; 8
3	Decide who the authors will be	7–8
4	Draft a working title and abstract	8–9
5	Decide on the basic form of the paper	9–10
6	Collect the material under the major headings chosen	10
7	Design tables, including their titles and footnotes; design or select illustrations and write titles and legends for them	10–15
8	Write for permission to reproduce any previously published tables, illustrations or other material that will be used	15–16
9	Write a topic outline and perhaps a sentence outline	16–19
10	Write, type or dictate a preliminary draft of the text quickly, to give it unity	20–25
11	Check completeness of the references assembled	25–26
12	Put the manuscript or typescript away for a few days	28
13	Re-examine the structure of the paper	29–30
14	Check the illustrations and tables and make the final versions	30–35
15	Re-read the references you cite and check your own accuracy in citing them; check for consistency, and reduce the number of abbreviations and footnotes	35–38
16	(Re)type the paper (= first draft)	38

Step		*Pages*
17	Correct the grammar and polish the style	38–45
18	Type several copies of the corrected paper (= second draft)	45
19	Ask for criticism from co-authors and friends	45
20	Make any necessary alterations	46
21	Compose a new title and abstract suitable for information retrieval, list the index terms and assemble the manuscript	46–52
22	Compile the reference list, cross-check references against the text, and ensure that all bibliographical details are correct	52–57
23	Retype (= penultimate version) and check typescript	57–58; 59–66
24	Obtain a final critical review from a senior colleague	58
25	Make any final corrections (final version)	67
26	Write a covering letter to the editor, enclosing copies of letters giving you permission to reproduce any previously published material or to cite unpublished work	67
27	Check that all parts of the paper are present, and post as many copies as specified to the editor	67–69
28	If the editor returns the paper, revise it as necessary, send it elsewhere, or abandon it	70–72
29	Correct the proofs	73–79

Appendix 2

Units of measure and their abbreviations

The Système International d'Unités (SI) is based on seven base units and two supplementary units. The base quantities, base units and their symbols are:

length	metre (m)
mass	kilogram (kg)
time	second (s)
current	ampere (A)
thermodynamic temperature	kelvin (K)
luminous intensity	candela (cd)
amount of substance	mole (mol)

The two supplementary units are:

plane angle	radian (rad)
solid angle	steradian (sr)

Strictly speaking, volumes should be expressed in this system in cubic metres (m^3), but the litre—now redefined as 1 dm^3—is more convenient in medicine and biology. Again, although molar concentrations should properly be given in moles per litre (mol/l), in practice molar (M) is accepted by many journals (though not by all).

Certain *derived* SI units have approved names and symbols. Their definitions in terms of the base units are shown in Table 1.

Certain prefixes (Table 2) are recommended for use with the SI units. In general, these represent multiples or submultiples of the form 10^{3n} or 10^{-3n}. In addition, the convenient centi, deci, deka, and hecto are allowed when the standard prefixes would yield cumbersome or conceptually absurd units (e.g. 15 cm^3 rather than 15 000 mm^3 or 0.000 015 m^3).

TABLE 1

Derived SI units

Physical quantity	*SI unit*	*Symbol or definition for unit*
force	newton (N)	kg m s^{-2}
pressure	pascal (Pa) (N/m^2)	kg m^{-1} s^{-2}
energy	joule (J)	kg m^2 s^{-2}
power	watt (W)	kg m^2 s^{-3}
electric charge	coulomb (C)	A s
electric potential	volt (V)	kg m^2 A^{-1} s^{-3}
electric capacitance	farad (F)	A^2 s^4 kg^{-1} m^{-2}
electric conductance	siemens (S)	A^2 s^3 kg^{-1} m^{-2}
electric inductance	henry (H)	kg m^2 A^{-2} s^{-2}
electric resistance	ohm (Ω)	kg m^2 A^{-2} s^{-3}
magnetic flux	weber (Wb)	kg m^2 A^{-1} s^{-2}
magnetic flux density	tesla (T)	kg A^{-1} s^{-2}
luminous flux	lumen (lm)	cd sr
illumination	lux (lx)	cd sr m^{-2}
frequency	hertz (Hz)	s^{-1} (cycles per second)
temperature interval	kelvin (K)	
Celsius temperature*	degree Celsius (°C)	(0°C = 273.15 K)
Celsius temperature interval*	degree Celsius	1 K

* Customary units that may continue to be used.

TABLE 2

Prefixes for SI units

10^{-18}	atto (a)	10^{-1}	deci (d)
10^{-15}	femto (f)	10	deka (da)
10^{-12}	pico (p)	10^2	hecto (h)
10^{-9}	nano (n)	10^3	kilo (k)
10^{-6}	micro (μ)	10^6	mega (M)
10^{-3}	milli (m)	10^9	giga (G)
10^{-2}	centi (c)	10^{12}	tera (T)

Never use two prefixes, but 'melt' them into one: 6 mμg = 6 ng; 4 μμl = 4 pl. When you raise the power of any unit that contains a prefix, there is no need to use parentheses: 2 km^2 is 2 $(km)^2$ or two square kilometres (2×10^6 m^2), *not* 2 $k(m^2)$ or 2×10^3 m^2. Similarly, 1 mm^3 is a cubic millimetre and 1 cm^3 a cubic centimetre.

The 'micron' is no longer acceptable; replace it by micrometre (μm). Similarly replace 'lambda' by microlitre (μl) and 'gamma' by microgram (μg). In these cases, the number will remain the same, but when you convert other non-SI units into SI units, you may need to apply a conversion factor (for example, 1 ångström = 10^{-1} nanometre; 1 dyne = 10^{-5} newton. For the necessary conversion factors see below (Table 3).

TABLE 3

Alphabetical list of abbreviations of units and prefixes

ampere: A
ångström: 10^{-1} nm
atmosphere, standard: 101 325 Pa
atomic mass unit (unified): u (1.660×10^{-27} kg)
atto: a (10^{-18})
bar: 10^{5} Pa
barn: 10^{-28} m^2
billion, if 10^9 (American usage): G
If 10^{12} (British usage): T
calorie: Use SI (cal_{IT} = 4.1868 J; cal (15°) = 4.1855 J; cal (thermochemical) = 4.184 J)
Calorie (medical) = kcal ≈ 4.2 kJ
candela: cd
candela steradian: cd sr (= lumen)
centrifugation, speed and duration: *g*-min
coulomb: C
count per minute: cpm or count/min
count per second: count/s
cubic centimetre: cm^3
cubic decimetre (= litre): dm^3
curie: Ci (3.7×10^{10} s^{-1})
cycle per second: Hz
dalton: *see* atomic mass unit
day: d (if absolutely necessary)
decibel: dB
degree: ° (plane angle)
degree Celsius: °C
degree Kelvin: does not exist. Use the kelvin (K)
disintegration per minute: dpm
disintegration per second: dps
dram: 1.117 g
dyne: 10^{-5} N
electronvolt: eV (0.160 219 aJ)
equivalents: eq or equiv (use moles if possible)
erg: 0.1 μJ
farad: F
foot: 0.305 m
foot-candle: 10.76 lux
gallon: Use metric: US, 3.785 litre; imperial, 4.546 litre
gamma: If microgram meant, use μg
gauss: 0.1 mT
giga: G
grain: 65 mg
gram, gramme: g
gram(me)-atom: g-atom
gram(me)-calorie: *see* calorie
gram(me)-centimetre: g·cm
gram(me)-molecule: mol
hecto: h
henry: H
hertz: Hz
horsepower: 745.7 W
hour: h
hundredweight: 51 kg
inch: 25.4 mm
joule: J (energy, including heat, work, nutrition)
katal: kat (proposed SI unit of enzyme activity)
kelvin: K
kilo: k
kilocalorie: 4.2 kJ
kilogram(me): kg
kilogram-force: 9.807 N
kilowatt hour: 3.6 MJ
knot: 1854 m/h
lambda: If microlitre meant, use μl
langley: 41.87 kJ m^{-2}
litre: l
lumen: lm
lux: lx

Table 3, *continued*

maxwell: 10^{-8} Wb
mega: M
megacycles per second: MHz
metre: m
mho: Use siemens, S
micro: μ
microlitre: μl (*not* λ)
micromicro: Substitute pico, p
micromilli: Substitute nano, n
micron: Use μm
mile: 1.609 km
mile, nautical: 1.854 km
milli: m
milliequivalent: Convert to millimolar, mM if possible
milligram percent: *see* p. 51
millimetre of mercury: mmHg; Give also SI equivalent (1 mmHg = 133.3 Pa)
millimicro: Substitute nano, n
millimicron: Substitute nanometre, nm
minute: min
minute (arc): (π/10 800) rad; ′
molal: Substitute mol/kg
molar (mole per litre): mol/l or M (note: M never means mole)
mole: mol
nano: n
nautical mile: 1.854 km
newton: N
normal (concentration): Use molar
ohm: Ω
ounce: 28.4 g
parts per million: parts/10^6
pascal: Pa
pico: p
poise: 10^{-1} Pa s
pound: 0.454 kg
pound per square inch: 6895 Pa
radian: rad
radiation absorbed dose: rad (10 mJ/kg) (do not abbreviate further)
radiation dose equivalent: rem
revolution per minute: rev/min
revolution per second: rev/s
roentgen: R (0.258 mCi/kg)
second: s
second (arc): ″
siemens: S
steradian: sr
stilb: cd/cm^2
stoke: 10^{-4} m^2 s^{-1}
Svedberg unit (10^{-13} s): S
Svedberg flotation unit: S_f
tera: T
tesla: T
ton: (2240 lb) 1016 kg; (2000 lb) 907 kg
tonne: Mg (10^3 kg)
torr: 133.3 Pa
unit: U or u
volt: V
watt: W
weber: Wb
week: wk
yard: 0.914 m
year: Avoid abbreviating. a (annus) is international symbol; use yr if abbreviation essential

Appendix 3

General abbreviations and symbols

Abbreviations given below are all acceptable; those marked with an asterisk must be defined at first mention in the text or in a table, preferably also in a 'list of abbreviations used' early in the paper. This list is compatible in most respects with abbreviations given by *Biological Abstracts* (1974); *Quantities, Units and Symbols* (Royal Society, Symbols Committee, 1971); and *Units, Symbols and Abbreviations* (Ellis [Royal Society of Medicine] 1972).

absorbance; absorbancy: *A*
acceleration due to gravity: *g* (9.81 m/s^2)
acidic dissociation constant: K_a
ad libitum: ad lib.
alternating current: a.c. or AC
and others: *et al.* (*with* the stop)
anno Domini (with dates): A.D.
ante meridiem (before noon): a.m. *Prefer* 24-hour clock: 0930 or 11:45 etc.
atomic weight: at. wt.; atomic mass preferred
bacille Calmette-Guérin: BCG
basal metabolic rate: BMR*
before present (with ages): B.P.*
blood pressure: BP*
boiling point: b.p.
centrifugation, speed and duration: *g*-min
cerebrospinal fluid: CSF
change per 10 K rise: Q_{10}
circa (about): ca.
compare: cf.
concentrated: concd
concentration: concn
constant, equilibrium: *K*
correlation coefficient: r (of observed sample), ρ (of hypothetical population)
date: prefer form 15 February 1974, abbreviated if necessary to 15 Feb 74 or 1974-02-15
degrees of freedom (statistics): df
density: ρ (mass); *d* (relative)
dextro: D

* Recommended abbreviation, but define at first mention.

dextrorotatory: (+)-
direct current: d.c. or DC
effective dose, median: ED_{50}
electrocardiogram: ECG*
electroencephalogram: EEG*
electron paramagnetic resonance: EPR or epr
electron spin resonance: ESR or esr
enzymes: in general, do not abbreviate; give Enzyme Commission number at first mention
equation: Eq.
equivalents: eq or equiv (use moles if possible)
erythrocyte sedimentation rate: ESR*
et alii: *et al.* (*with* the stop)
et cetera: etc. Use rarely and only when you have specific things in mind
experiment: expt
experimental: exptl
exponential: exp
extinction [$\log_{10}(I_o/I)$]: *E*
extinction coefficient, molar: ε
female: F; ♀
figure, figures: Fig., Figs.
first filial generation: F1
focus-forming units: FFU*
for example: e.g. (exempli gratia)
freezing point: f.p.
gas–liquid chromatography: GLC*
gravity, unit of acceleration of: *g* (9.81 m/s^2)
greater than: >
half-life: $T_{1/2}$
hydrogen ion concentration, negative log of: pH
infective dose, median: ID_{50}*
infrared: i.r. or IR
inner diameter: i.d. or ID
inside diameter: i.d. or ID
intelligence quotient: IQ
international unit: i.u. or IU
intraarterial(ly): i.a.*
intracutaneous(ly): i.c.*
intradermal(ly): i.d.*
intramuscular(ly): i.m.*
intraperitoneal(ly): i.p.*
intravenous(ly): i.v.*
isoelectric point: pI
laevo (configuration): L
laevorotatory: (–)-
less than: <
lethal dose, median: LD_{50}
logarithm: log (base 10 understood); ln (base *e*); $\log_n$ (any base specified by *n*)
longitude: long.
male: M; ♂
maximum, maximal: max
median effective dose: ED_{50}*
median infective dose: ID_{50}*
median lethal dose: LD_{50}*
melting point: m.p.
meta-: *m-*
metabolic quotient: Q
minimum, minimal: min
molar extinction coefficient: ε
molecular optical rotation: $[M]_D^{20}$
molecular weight: mol. wt.; M
normal temperature and pressure: Use standard temperature and pressure
not determined: ND*
not significant: NS*

* Recommended abbreviation, but define at first mention.

nuclear magnetic resonance: NMR or nmr
number: no.
observed, observation: obs.
opere citato (in the work cited): op. cit.
optical rotation, molecular: $[M]_D^{20}$
optical rotation, specific: $[\alpha]_D^{20}$
optical rotatory dispersion: ORD*
ortho-: *o-*
osmotic pressure: Π
outside diameter: o.d. or OD
para-: *p-*
partial specific volume: $\bar{v}$
parts per million: parts/10^6
per cent (Brit), percent (US): %
per thousand: /10^3. Avoid $^0/_{00}$
phenylketonuria: PKU*
plaque-forming units: PFU*
plus or minus: ±
post meridiem (after noon): p.m. *Prefer* 24-hour clock: 1530 or 21:40, etc.
potential difference: p.d.
precipitate: ppt.
probability: P or *P*
proton magnetic resonance: PMR or pmr
radioisotopes: ^{14}C, ^3H, etc.; carbon-14 permissible
rapid eye movement: REM*
red blood corpuscle: *prefer* erythrocyte
refractive index: *n*; n_D^{20}
respiratory quotient: RQ*
retardation factor (chromatography): R_f
reticuloendothelial system: RES*
rhesus (blood factor): Rh
sedimentation coefficient: *s*
simian virus: SV*
specific activity: sp. act.
specific gravity: sp. gr.
standard deviation: SD
standard error (of sampling): SE
standard error of the mean: SEM
standard temperature and pressure: s.t.p.
Student's t: *t* (with number of degrees of freedom)
subcutaneous(ly): s.c.*
temperature: temp; *T* (absolute, in equations); *t* (empirical)
temperature-sensitive: ts*
that is: i.e. (id est)
thin-layer chromatography: TLC*
ultraviolet: u.v. or UV
uniformly labelled: U-
unit: U or u
vapour density: v.d.*
vapour pressure: v.p.*
variance ratio: *F* (statistics)
versus: vs.
volume: vol
volume ratio: v/v (two components only; otherwise 'by vol.')
wavelength: λ
white blood corpuscle: *Prefer* leucocyte
weight per volume: wt/vol; *Prefer* mass/vol
weight ratio: wt/wt

* Recommended abbreviation, but define at first mention.

Appendix 4

Abbreviations in biochemistry and taxonomy

Abbreviations in biochemistry and taxonomy are given here because these two disciplines have applications in much of biology and because abbreviations tend to proliferate in both of them. More advanced guidance is given in Sober (1968) and *Biological Abstracts* (1974) respectively.

Abbreviations given below are all acceptable; those marked with an asterisk must be defined at first mention in the text or in a table, preferably also in a 'list of abbreviations used' early in the paper.

acetate: OAc
acetyl: Ac
N-acetyl: NAc
acetyl coenzyme A: CoASAc; acetyl-CoA
N-acetylglucosamine: GalNAc
adenosine: Ado; A (in sequences)
adenosine 5′-phosphates: AMP, ADP, ATP
adenosine-ribosylthymine: A-T
adrenocorticotropic hormone, adrenocorticotropin: ACTH
alanine residue: Ala
amino acids: use 3-letter symbols only for aminoacyl residues in sequences, not for free amino acids. See under each amino acid for its symbol
arginine residue: Arg
asparagine residue: Asn
asparagine or aspartic acid residue: Asx
aspartic acid residue: Asp
complement and its components: C, C1, C2, etc*
cultivar: cv.
cysteine residue: Cys
cystine residue: Cys; Cys
cytidine: Cyd; C (in sequences)
cytidine 5′-phosphates: CMP, CDP, CTP
deoxy: d (in nucleotides)

* Recommended abbreviation, but define at first mention.

deoxyribonuclease: DNase
deoxyribonucleic acid: DNA
deoxyribosylthymine: dThd; dT (in sequences)
dextro: D
dextrorotatory: (+)-
O-(diethylaminoethyl)cellulose: DEAE-cellulose*
3,4-dihydroxyphenylethylamine: dopamine
3,4-dihydroxyphenylalanine: dopa
dimethylaminonaphthalene-*S*-sulphonyl: dansyl, Dns*
2,4-dinitrophenyl: DNP* (with trivial names); Dnp (with amino acid symbols)
diphosphopyridine nucleotide, oxidized form: NAD*
diphosphopyridine nucleotide, reduced form: NADH
doubtful name: nom. dub.
emendation, emended: emend.
enzymes: in general, do not abbreviate; give Enzyme Commission number at first mention
ethylenediaminetetraacetic acid or acetate: EDTA
familia nova: fam. nov.
flavin-adenine dinucleotide: FAD
flavin-adenine dinucleotide, reduced form: $FADH_2$
flavin mononucleotide: FMN
follicle-stimulating hormone: FSH*
forma: f.
forma specialis: f.sp.
formylmethionyl: fMet
fructose residue: Fru
fucose residue: Fuc
furanose: $_f$ (as suffix)
galactosamine: GalN
galactose residue: Gal
genus novum: gen. nov.
gluconic acid: GlcA
glucosamine: GlcN
glucose: Glc (residue); G (in some combinations)
glucose 6-phosphate: Glc-6-P; G-6-P
glucuronic acid: GlcUA
glutamic acid residue: Glu
glutamic acid or glutamine residue: Glx
glutamine residue: Gln
glutathione, oxidized: GSSG
glutathione, reduced: GSH
growth hormone: GH*
guanosine: Guo; G (in sequences)
guanosine 5′-phosphates: GMP, GDP, GTP
haemoglobin: Hb
histidine residue: His
hydroxylysine residue: Hyl
hydroxyproline residue: Hyp
immunoglobulins: IgA, IgD, IgE, IgG, IgM
incertae sedis (uncertain position): inc. sed.
inorganic phosphate: P_i
inosine: Ino; I (in sequences)
inosine 5′-phosphates: IMP, IDP, ITP
invalid name: nom. nud.
isoleucine residue: Ile
lactose residue: Lac
laevo (configuration): L
laevorotatory: (–)-

---

* Recommended abbreviation, but define at first mention.

leucine residue: Leu
luteinizing hormone: LH*
lysergic acid diethylamide: LSD*
lysine residue: Lys
maltose residue: Mal
mannose residue: Man
melanocyte-stimulating hormone: MSH*
messenger ribonucleic acid: mRNA*
meta-: *m*-
methionine residue: Met
methyl: Me
Michaelis constant: K_m or K_m
minimum, minimal: min
molar extinction coefficient: ε
morpha (form): m.
myoglobin: Mb
myoglobin, carboxy: MbCO
myoglobin, oxy: MbO_2
N-acetylgalactosamine: GalNAc
N-acetylglucosamine: GlcNAc
new combination: comb. nov.
new family: fam. nov.
new genus: gen. nov.
new name: nom. nov.
new species: sp. nov.
new subspecies: ssp. nov.
new variety: var. nov.
nicotinamide-adenine dinucleotide: NAD
nicotinamide-adenine dinucleotide, oxidized form: NAD*
nicotinamide-adenine dinucleotide, reduced form: NADH
nicotinamide-adenine dinucleotide phosphate, oxidized form: NADP*
nicotinamide-adenine dinucleotide phosphate, reduced form: NADPH
nicotinamide mononucleotide: NMN
nomen conservandum: nom. cons.
nomen dubium: nom. dub.
nomen nudum: nom. nud.
nomen rejiciendum: nom. rej.
nuclear RNA: nRNA*
nucleoside, unspecified: N (*not* X)
optical density: OD (for suspensions); substitute absorbance (*A*) for solutions
ornithine residue: Orn
ortho-: *o*-
orthophosphate: P_i
oxyhaemoglobin: HbO_2
oxymyoglobin: MbO_2
para-: *p*-
parts per million: ppm
perchloric acid: $HClO_4$ *not* PCA
phenyl: Ph*
phenylalanine residue: Phe
phosphate, inorganic: P_i
phosphate other than inorganic: *P* (sometimes *p*)
phosphocreatine: *P*-creatine
phosphoenolpyruvate: *P*-enolpyruvate
preoccupied: preocc.
proline residue: Pro
pseudouridine: Ψrd; Ψ (in sequences)
pseudouridine 5′-phosphates: ΨMP, ΨDP, ΨTP
purine nucleoside, unspecified: Puo; R (in sequences)

* Recommended abbreviation, but define at first mention,

pyranose: $_p$ (as suffix)
pyrophosphate, inorganic: PP_i
rejected name: nom. rej.
retained name: nom. cons.
riboflavin 5′-phosphate: FMN
ribonuclease: RNase
ribonucleic acid: RNA
ribose residue: Rib
ribosomal RNA: rRNA*
ribosyl (where needed before nucleotide symbol): r
ribosylnicotinamide residue: Nir
ribosylthymine: Thd; T (in sequences)
ribosyluracil: Ψrd; Ψ (in sequences)
sensu lato: s.l.
sensu stricto: s.s.
serine residue: Ser
soluble RNA: Substitute tRNA
sorbose residue: Sor
species: sp., spp.
species nova: sp. nov.
subspecies: ssp., sspp. (plural)
subspecies nova: ssp. nov.
sucrose residue: Suc
sulphydryl: SH
thiol: SH
threonine residue: Thr
thymidine: dThd; dT (in sequences)
thymidine 5′-phosphates: dTMP, dTDP, dTTP
thyroid-stimulating hormone: TSH*
tobacco mosaic virus: TMV*
transfer RNA: tRNA
trichloroacetic acid: TCA*; Prefer CCl_3COOH
1,1,1-trichloro-2,2-bis(*p*-chlorophenylethane): DDT*
O-(triethylaminoethyl)cellulose: TEAE-cellulose*
triphosphopyridine nucleotide, oxidized form: NADP*
triphosphopyridine nucleotide, reduced form: NADPH
tris(hydroxymethyl)aminomethane: Tris*
tryptophan residue: Trp (*not* Try)
tyrosine residue: Tyr
uridine: Urd; U (in sequences)
uridine 5′-phosphates: UMP, UDP, UTP
valine residue: Val
variety: var.
variety, new: var. nov.
xanthosine: Xao; X (in sequences)
xanthosine 5′-phosphates: XMP, XDP, XTP
xylose residue: Xyl

---

* Recommended abbreviation, but define at first mention.

Appendix 5

Expressions to avoid

Readers are advised to work through the whole list to get the flavour of the suggestions rather than use it as a reference tool, since alphabetization is so arbitrary. Do *not* regard the list as a dictionary: the word or phrase in the second column is not by any means always an exact equivalent of that in the first.

Avoid	*Usually prefer*
aliquot	portion, sample [an aliquot is an integral subdivision of the whole: 2 ml can be described as an aliquot of 10 ml, but 3 ml cannot]
alternate (adjective)	[see pp. 43–44: you probably need 'alternative']
anticipate	expect
approximately	about
are of the same opinion	agree
area	*see* 'woolly words' below
as already stated	[omit]
as can be seen from Fig. 1, growth is more rapid	growth is more rapid (Fig. 1), *or* Fig. 1 shows that growth is more rapid
as far as our own observations are concerned, they show	our/my observations show
as far as this species is concerned, it	this species is
as follows:–	[omit 'as follows' and the dash; the colon is sufficient]
as for these experiments, they are	these experiments are

Avoid	*Usually prefer*
as of now	now, from now on
as shown in Fig. 11	Fig. 11 shows that
as regards this species, it	this species is
assist(-ance)	help, aid
at some future time	later
at the present moment, at the present moment in time, at this time	now
author(s), the	I / we
bright red in colour	bright red
case	patient
[colourless past participles: accomplished, achieved, attained, carried out, conducted, done, effected, experienced, facilitated, given, implemented, indicated, involved, made, obtained, occurred, performed, proceeded, produced, required]	[avoid; use a meaningful verb instead]
commence	begin, start
comparatively	[avoid, unless you are making a real comparison of one item with another]
concerning this effect, it may be borne in mind that	[omit]
conducted inoculation experiments on	inoculated
considerable amount of	much
considerable number/proportion of	many, most
contemporaneous in age	contemporaneous; the same age
created the possibility	made possible; enabled (a person); allowed (an action)
data	facts, results, observations
decreased number of	fewer, less
decreased relative to	less than, lower than
demonstrate	show
due to the fact that	because
during the time that	while

Avoid	*Usually prefer*
elevated	raised, higher, more
employ	use
encountered	met
encountered frequently (e.g. 'this effect was encountered frequently')	common ('this effect was common')
equally as well	equally well
exhibit; X exhibits good stretch properties	show; X stretches well
fairly	[omit]
fewer in number	fewer
following (the operation)	after (the operation)
for the reason that	because, since
from the standpoint of	according to
goes under the name of	is called
greater/higher number of	more
hospitalize	admit to hospital
hydroxylation reaction	hydroxylation [-tion ending denotes a reaction]
if and when	[use one of these prepositions alone]
if conditions are such that	if
in a considerable number of cases	often
in all cases	always
in connection with	about, for
increased relative to	more than
in few cases	sometimes, rarely
in excess of	more than, above
in order to	to
in regard to in relation to in respect of in terms of in the case of in the context of	[use the appropriate preposition, e.g. in, for, about, with]
in the event that	if
in the present communication	here; in this paper

Avoid	*Usually prefer*
in this connection the statement may be made that	[omit]
in view of the fact that	since, because
it has long been known that	[omit]
it is apparent, therefore, that	hence, therefore
it is of interest to note that	[omit]
it is often the case that	often
it is possible (probable) that	possibly (probably)
it is this that	this
it may, however, be noted that	nevertheless, but
it may be said that	possibly
it seems to the present writer	I think
it will be seen upon examination of Table 5 that	Table 5 shows that
it would thus appear that	apparently
large number(s) of	many
large proportion of	much, most
lazy in character	lazy
lesser extent, degree	less
level	concentration, content
literature, e.g. It is reported in the literature that	others have reported that
made a count	counted
majority of	most
mental patients	patients with mental disorders
moment in time	time
much	[omit]
multiple	several, different
number of	several, some
of large size	large
of such strength that	so strong that
on the basis of	from, by, because
owing to the fact that	because
parameter	index, criterion, factor, characteristic, measure, value, variable
pertaining to	on, about
prior to; prior to that time	before; before that
proportion of	some

Avoid	*Usually prefer*
quite	[omit]
rather	[omit]
regime	regimen
relatively	[avoid, unless you are calculating a real quotient]
relative to ('This letter is relative to')	about
respectively	[avoid, unless you are sure you are using the word correctly, as in 'Absorption measurements for cadmium vapour at 50, 70, and 100 nm give cross-sections of 0.7, 0.9, and 1.2 fm², respectively.']
reveal	show
sacrifice (experimental animals)	kill
serves the function of being	is
significantly	much, appreciably, definitely [but correct and desirable if statistical tests of significance have been made]
similar in every detail	the same
small numbers of	few
sophisticated	advanced, new, expensive
species in which the hairs are lacking	hairless species
square in shape, square-shaped	square
sufficient number of	enough
subsequent to	after
terminate	end
the test in question	this test
the tests have not as yet	the tests have not
the treatment having been performed,	after treatment
there can be little doubt that this is	this is probably
there is, there are	[often unnecessary: e.g. change 'There is much work being done on . . .', to 'Much work is being done on . . .']
there will always be a miscellany of quality in terms of illustrations	the quality of illustrations will always vary

Avoid	*Usually prefer*
they are both alike (similar)	they are alike (similar)
throughout the entire area	throughout the area
throughout the whole of the experiment	throughout the experiment
two equal halves	halves
upon	on
using	[test whether it is an unattached participle; if so, try 'by', 'with', 'by means of']
utilize, utilization	use
very	[omit]
while	although
with reference to	about
with regard to	in, to
[woolly words: area, character, conditions, field, level, nature, problem, process, situation, structure, system]	[avoid: change to more precise words]

Bibliography

References Cited, including American, British and International Standards

AFNOR (1973) *Présentation des Articles de Périodiques* [Draft international standard], ISO/TC 46/GT 7, AFNOR, Paris, NF Z 41–003

American National Standards Institute (1969) *Abbreviation of Titles for Periodicals*, ANSI Z39.5–1969, American National Standards Institute, New York, N.Y.

American National Standards Institute (1971) *American National Standard for Writing Abstracts*, Z39.14–1971, ANSI, New York, N.Y.

American National Standards Institute (1972) *American National Standard for the Preparation of Scientific Papers for Written or Oral Presentation*, Z39.16–1972, ANSI, New York, N.Y.

Biological Abstracts (1974) Guide to the preparation of citations and abstracts, *Biological Abstracts*, xxv-xxvii

BIOSIS (1974) *List of Serials*, BioSciences Information Service of Biological Abstracts, Philadelphia, Pa.

BIOSIS, Chemical Abstracts Service & Engineering Index, Inc. (1974) *Bibliographic Guide for Editors and Authors*, Biological Abstracts, Philadelphia; Chemical Abstracts, Columbus, Ohio; Engineering Index, Inc., New York, N.Y.

British Standards Institution (1958) *Transliteration of Cyrillic and Greek characters*, BS 2979, BSI, London

British Standards Institution (1970) *The Abbreviation of Titles of Periodicals*, BS 4148: Part 1, British Standards Institution, London

British Standards Institution (1972) *Draft British Standard Recommendations for Publishers' and Printers' Style Manuals*, BSI, London

British Standards Institution (1973) *Draft British Standard Specification for Copy Preparation and Proof Correction (Revision of BS 1219)*, BSI, London

CAREY, G. V. (1971) *Mind the Stop: A Brief Guide to Punctuation, with a Note on Proof-Correction*, Penguin Books, Harmondsworth, Middlesex

Chemical Abstracts (1969) *ACCESS: Key to the Source Literature of the Chemical Sciences*, American Chemical Society, Columbus, Ohio

Commission on Biochemical Nomenclature (1973) *Enzyme Nomenclature*, Elsevier, Amsterdam

Council of Biology Editors (1972) *CBE Style Manual*, 3rd edn., American Institute of Biological Sciences, Washington, D.C.

CSIRO (1973) *Guide to Authors* [Draft of *CSIRO Style Guide* (1975, in press) for those submitting papers to the Australian journals of scientific research.] Commonwealth Scientific and Industrial Research Organization, Melbourne

ELLIS, G. (ed.) (1972) *Units, Symbols and Abbreviations*, 2nd edn., Royal Society of Medicine, London

FOWLER, H. W. (revised by Gowers, Sir Ernest) (1965) *A Dictionary of Modern English Usage*, 2nd edn., Oxford University Press, London

GOWERS, Sir Ernest (revised by Fraser, Sir Bruce) (1973) *The Complete Plain Words*, 2nd edn., Her Majesty's Stationery Office, London

HORNBY, A. S. (1954) *A Guide to Patterns and Usage in English*, Oxford University Press, London

Index Medicus, National Library of Medicine, Bethesda, Maryland

International List of Periodical Title Word Abbreviations (1970) Prepared for the UNISIST/ICSU-AB Working Group on Bibliographic Descriptions, ICSU-AB Secretariat, Paris; Chemical Abstracts Service, Columbus, Ohio

MARLER, E. E. J. (1973) *Pharmacological and Chemical Synonyms*, 5th edn., Excerpta Medica, Amsterdam

Royal Society (1974) *General Notes on the Preparation of Scientific Papers*, Royal Society, London

Royal Society, Symbols Committee (1971) *Quantities, Units and Symbols*, Royal Society, London

SOBER, H. A. (ed.) (1968) *Handbook of Biochemistry*, Chemical Rubber Company, Cleveland, Ohio

STRUNK, W., JR, & WHITE, E. B. (1972) *The Elements of Style*, 2nd edn., Macmillan, Riverside, N.J.; Collier-Macmillan, London

TICHY, H. J. (1967) *Effective Writing for Engineers, Managers, Scientists*, Wiley, New York and London

TRELEASE, S. F. (1969) *How to Write Scientific and Technical Papers*, M.I.T. Press, Cambridge, Mass. and London, England

Ulrich's International Periodicals Directory. Fifteenth edn., 1973–1974, Bowker, New York and London

WOODFORD, F. P. (ed.) (1968) *Scientific Writing for Graduate Students: A Manual on the Teaching of Scientific Writing*, Rockefeller University Press, New York; Macmillan, London

World List (1963–1965) *World List of Scientific Periodicals Published in the Years 1900–1960* 4th edn. (Brown, P. & Stratton, G.B., eds.), Butterworth, London

Additional Reading

American Chemical Society (1967) *Handbook for Authors*, American Chemical Society Publications, Washington, D.C.

American Institute of Physics (1973) *Style Manual*, American Institute of Physics, New York, N.Y.

American Medical Association (1971) *Style Book and Editorial Manual*, 5th edn., Scientific Publications Division, American Medical Association, Chicago, Ill.

BAKER, J. R. (1955) English style in scientific papers. *Nature* (*Lond.*) *176*, 851–852

BARABAS, A. & CALNAN, J. (1973) *Writing Medical Papers: a Practical Guide*, Heinemann Medical Books, London

COCHRAN, W., FENNER, P. & HILL, M. (eds.) (1973) *Geowriting: A Guide to Writing, Editing and Printing in Earth Sciences*, American Geological Institute, Washington, D.C. (GEGS Programs Publication 17)

FOLLETT, W. (1970) *Modern American Usage: a Guide*, Grosset & Dunlap, New York, N.Y.

IUB (International Union of Biochemistry) – CEBJ (Commission of Editors of Biochemical Journals) (1973). The citation of bibliographic references in biochemical journals. *Biochem. J. 135*, 1–3

PARTRIDGE, E. (1970) *Usage and Abusage*, Penguin Books, Harmondsworth, Middlesex, and Baltimore, Md. [First published by Hamish Hamilton, London, 1946]

QUILLER-COUCH, A. (1916) *The Art of Writing*, Cambridge University Press

THORNE, C. (1970) *Better Medical Writing*, Pitman Medical, London

TURNER, R. P. (1964) *Grammar Review for Technical Writers*, Holt, Rinehart & Winston, New York, N.Y.

University of Chicago Press (1969) *A Manual of Style*, 12th edn., University of Chicago Press, Chicago, Ill.

WEST, M. & KIMBER, P. F. (1958) *A Deskbook of Correct English*, Longmans Green, London

Index

→

This pocket and the one inside the front cover are intended to hold the two series of guidelines—one series of booklets for special language groups and the other for the different disciplines of science—that are being prepared and issued separately from this volume. (Please note that these booklets will *not* be published by Associated Scientific Publishers; further information about the two series can be obtained from the Secretary of ELSE: Dr. J. R. Metcalfe, Commonwealth Institute of Entomology, 56, Queen's Gate, London, SW7 5JR, UK.)